3

WATER RESOURCE PLANNING, DEVELOPMENT AND MANAGEMENT

RURAL WATER SYSTEMS

CHALLENGES AND DRINKING WATER NEEDS

WATER RESOURCE PLANNING, DEVELOPMENT AND MANAGEMENT

Additional books in this series can be found on Nova's website
under the Series tab.

Additional e-books in this series can be found on Nova's website
under the e-book tab.

WATER RESOURCE PLANNING, DEVELOPMENT AND MANAGEMENT

RURAL WATER SYSTEMS

CHALLENGES AND DRINKING WATER NEEDS

GLORIA SIMMONS
EDITOR

New York

Library of Congress Cataloging-in-Publication Data

ISBN: 978-1-63483-240-3

Published by Nova Science Publishers, Inc. † New York

CONTENTS

PREFACE

According to the Census Bureau, approximately 27 percent of the U.S. population lives in a rural area. The smallest water systems account for 77 percent of all systems, and many water systems are not in compliance with drinking water regulations. Seven federal agencies provide funding or technical assistance to rural communities in developing drinking water and wastewater systems and complying with federal regulations. The nation faces costly upgrades to aging and deteriorating drinking water and wastewater infrastructure. Many rural communities face significant challenges in financing the costs of replacing or upgrading aging and obsolete drinking water and wastewater infrastructure. This book focuses on the federal agencies that provide funding or technical assistance to rural communities and fiscal year 2014 funding; and issues identified that affect rural communities' ability to obtain funding for water and wastewater infrastructure. Furthermore, this book summarizes the Safe Drinking Water Act (SDWA) and its major programs and regulatory requirements.

In: Rural Water Systems
Editor: Gloria Simmons

ISBN: 978-1-63483-240-3

Chapter 1

BACKGROUND MEMO FOR THE HEARING ON "THE NEEDS OF DRINKING WATER SYSTEMS IN RURAL AND SMALLER COMMUNITIES"*

Staff of the House Committee on Energy and Commerce

TO :	Members, Subcommittee on Environment and the Economy
FROM:	Committee Majority Staff
RE:	Hearing entitled "The Needs of Drinking Water Systems in Rural and Smaller Communities"

II. BACKGROUND

According to the Census Bureau, approximately 27 percent of the U.S. population lives in a rural area.[1] The smallest water systems (serving fewer than 3,300 persons, many serving small clusters of homes) account for 77 percent of all systems, and many water systems are not in compliance with drinking water regulations.[2]

---

* This is an edited, reformatted and augmented version of a document prepared for a February 27, 2015 hearing of the House Committee on Energy and Commerce, Subcommittee on Environment and the Economy.

Many of these smaller and rural communities – with populations of 10,000 or less – face significant challenges in replacing or upgrading aging and obsolete drinking water and wastewater infrastructure.[3] For example, as regulations become more stringent, these communities may need to upgrade basic wastewater systems with more sophisticated equipment or more expensive filtration equipment to achieve regulatory compliance.[4] The Congressional Research Service cites State surveys suggesting funding needs in these areas may exceed $88 billion.[5]

While these small and rural customers pay for drinking water and wastewater operations-related and infrastructure costs through the rates charged by their systems, but unlike water systems in larger urban markets, these communities do not have the rate base to fund some projects and still maintain affordable rates. Many small and rural communities have limited access to financial markets, restricting their ability to issue bonds to raise capital.[6] As a result, these communities depend heavily on Federal and State grants and subsidized loan programs to finance their needs.

While several Federal programs assist rural communities in meeting water quality requirements – the Environmental Protection Agency (EPA) and the U.S. Department of Agriculture (USDA) oversee the major Federally funded drinking water and wastewater infrastructure assistance programs. The subject of today's hearing will be the Safe Drinking Water Act (SDWA) issues related to these small and rural drinking water utilities.

U.S. EPA Activities

Under the SDWA, small and rural drinking water supply systems are subject to a number of drinking water regulations issued by EPA. These requirements include system monitoring, treatment to remove certain contaminants, and reporting. EPA estimates that compliance with the regulations already promulgated will provide millions of people protection from numerous industrial chemicals, microbes, and other contaminants in public water supplies. However, to comply, many cities and towns must invest in capital equipment, operation and maintenance, and increased staff technical capacity. Recent regulations with particularly costly implications for small towns include water filtration, lead control, arsenic control, and inorganic and organic contaminant control.[7]

Moreover, new regulations are being developed on strontium and perchlorate that are likely to impose additional compliance burdens on the smaller and rural drinking water systems within the next few years. Funding existing burdens and these new requirements are a growing concern to water suppliers.

SDWA section 1452 provides EPA authority to grant funding to States to administer their own Drinking Water State Revolving Funds (SRF), which annually provide financial assistance to communities of all sizes for expenditures (not including monitoring, operation, and maintenance expenditures) that EPA has determined will facilitate a system's compliance with drinking water regulations or otherwise further the SDWA's health protection objectives. In many instances, this means financing drinking water treatment plants, distribution and storage infrastructure, and source projects.[8]

In addition, because many of these systems lack resources or access to certain professionals, SDWA section 1442(e) authorizes EPA to provide technical assistance to "small public water systems" to enable those systems to achieve and maintain compliance with applicable Federal drinking water regulations, including through circuit-rider and multi-State regional technical assistance programs, training, and preliminary engineering evaluations. As an example, in 2012, the Government Accountability Office found that preparing additional engineering reports could cost from $5,000 to $50,000 and that the cost of a typical environmental analysis could cost between $500 and $15,000.[9]

Related Federal or Relevant Programs

Besides EPA, other Federal agencies administer programs that assist rural communities in developing water systems, including the U.S. Department of Agriculture (USDA), the Department of Housing and Urban Development (HUD), and the Economic Development Administration (EDA).

USDA's Rural Utilities Service (RUS), one of three agencies under USDA Rural Development, administers the Water and Waste Disposal program. This program, where USDA State offices work with State and local governments, Indian tribes, and private and community organizations to prepare a strategic plan for delivering assistance, provides funding (loans and grants) for both drinking water and wastewater projects in low-income rural communities of 10,000 or less who have been denied credit through normal commercial channels.[10] USDA prefers making loans, but grants are made to reduce average annual user charges. In recent years, approximately 65 percent of loan funds and 57 percent of grant funds have been obligated to drinking water projects.

RUS also is authorized to help rural residents in an emergency situation where a significant decline in quantity or quality of drinking water exists or is imminent and funds are needed to obtain adequate quantities of water that meet standards of the Safe Drinking Water Act. Funding these emergency grants is provided through reservation of 3 percent to 5 percent of appropriated RUS water and waste disposal grant funds.[11]

HUD and EDA offer assistance through programs geared toward economic development. Under HUD's Community Development Block Grant (CDBG) program, water needs compete with many other public activities and are estimated to account for 10 percent to 20 percent of CDBG obligations.[12]

EDA provides project grants for construction of water systems in economically distressed areas. Economic development grants can be used for a wide range of purposes and frequently have a sewer or water supply element.[13]

The Rural Community Assistance Partnership (RCAP) works with small, rural communities to build, maintain, or expand their water infrastructure through onsite, technical assistance; training in financial, managerial, and operational areas; and educational resources. RCAP is a national network of nonprofit organizations working with rural and small communities throughout the United States and receives funding from the U.S. government,.[14]

End Notes

[1] http://www.crs.gov/pages/Reports.aspx?PRODCODE=98-64&Source= search

[2] Ibid. 3 http://gao.gov/products/GAO-13-279SP#mt=e-report&f=/ereport/GAO-13- 279SP/data_center/Science_and_the_environment/14._Rural_Water_Infrastructure&ft=0

[4] Ibid.

[5] http://www.crs.gov/pages/Reports.aspx?PRODCODE=98-64&Source=search

[6] Ibid.

[7] http://www.crs.gov/pages/Reports.aspx?PRODCODE=98-64&Source=search

[8] SDWA §1452

[9] http://gao.gov/products/GAO-13-279SP#mt=e-report&f=/ereport/GAO-13- 279SP/data_center/Science_and_the_environment/14._Rural_Water_Infrastructure&ft=1

[10] Ibid.

[11] Ibid.

[12] http://www.crs.gov/pages/Reports.aspx?PRODCODE=98-64&Source=search#_Toc403401920

[13] Ibid.

[14] http://www.rcap.org/about

In: Rural Water Systems
Editor: Gloria Simmons

ISBN: 978-1-63483-240-3

Chapter 2

RURAL WATER INFRASTRUCTURE: FEDERAL AGENCIES PROVIDE FUNDING BUT COULD INCREASE COORDINATION TO HELP COMMUNITIES. STATEMENT OF ALFREDO GOMEZ, DIRECTOR, NATURAL RESOURCES AND ENVIRONMENT TEAM, GOVERNMENT ACCOUNTABILITY OFFICE. HEARING ON "THE NEEDS OF DRINKING WATER SYSTEMS IN RURAL AND SMALLER COMMUNITIES"*

WHY GAO DID THIS STUDY

The nation faces costly upgrades to aging and deteriorating drinking water and wastewater infrastructure. Many rural communities face significant challenges in financing the costs of replacing or upgrading aging and obsolete drinking water and wastewater infrastructure. The costs of replacing

* This is an edited, reformatted and augmented version of a report, GAO-15-450T, issued by the U.S. Government Accountability Office. This report was originally presented as testimony February 27, 2015 before the House Committee on Energy and Commerce, Subcommittee on Environment and the Economy.

infrastructure in these communities are estimated by federal agencies to be over $140 billion in the coming decades.

A number of federal agencies provide funding and technical assistance to rural communities for infrastructure development. GAO has previously reviewed these federal agencies and the funding they provide rural communities. The agencies define rural differently, according to their individual authorizations and guidelines.

This testimony is based on reports GAO issued from September 2007 through October 2012, with updated information, as appropriate, through February 2015. It focuses on (1) the federal agencies that provide funding or technical assistance to rural communities and fiscal year 2014 funding and (2) issues identified in GAO's work that affect rural communities' ability to obtain funding for water and wastewater infrastructure.

GAO made recommendations in its reports to improve federal agencies' coordination of funding programs for rural communities. EPA and USDA generally concurred with the recommendations, have taken action on some, and are beginning to take action on others.

WHAT GAO FOUND

Seven federal agencies provide funding or technical assistance to rural communities in developing drinking water and wastewater systems and complying with federal regulations. The Environmental Protection Agency's (EPA) Drinking Water and Clean Water Revolving Funds (SRF) are the largest source of funding and assistance receiving $907 million and $1.45 billion respectively in fiscal year 2014, some of which goes to rural communities. The Department of Agriculture's (USDA) Rural Utilities Service provides the next largest source of funding at $485 million in fiscal year 2014. The other five federal agencies that provide funding or technical assistance to rural communities are the Department of Housing and Urban Development; Department of Health and Human Service's Indian Health Service; Department of Commerce's Economic Development Administration; U.S. Army Corps of Engineers; and Department of the Interior's Bureau of Reclamation.

GAO's work on rural water infrastructure funding found the following issues that affect the ability of rural communities' to fund water and wastewater infrastructure.

- Communities typically paid for drinking water and wastewater infrastructure through the rates charged to users of the drinking water and wastewater systems. In some cases, however, these communities did not have the number of users needed to share the cost of major infrastructure projects while maintaining affordable user rates. As a result, they depended heavily on federal and state grants and subsidized loan programs.
- Some rural communities did not have technical expertise and had to hire consultants and engineers to help design water or wastewater projects and complete the technical documents necessary to apply for funding. This included developing preliminary engineering plans and environmental documents. Some federal and state programs pay for technical service providers that communities can use to help them design and finance their projects, and apply for funding.
- Rural communities faced potentially duplicative application requirements when applying to multiple state or federal programs, making it more costly and time-consuming to complete the application process. For example, engineers GAO interviewed estimated that preparing additional preliminary engineering work could cost anywhere from $5,000 to $50,000 and that the cost of an additional environmental analysis could add as little as $500 to a community's costs or as much as $15,000. As of February 2015, EPA, USDA, and several of the other federal agencies had taken steps to improve coordination of funding and assistance to rural communities. Most notably, the agencies developed a standard engineering report that communities can use to apply for funding from multiple agencies.

* * *

Chairman Shimkus, Ranking Member Tonko, and Members of the Subcommittee:

I am pleased to be here today as you consider the infrastructure needs facing the nation's rural drinking water and wastewater systems. As you know, the nation's water utilities face the task and costs of upgrading aging and deteriorating infrastructure in both drinking water plants and wastewater treatment systems, including collection systems, treatment plants, and distribution systems. The Environmental Protection Agency (EPA) estimates that the funding needs for drinking water infrastructure overall total $384.2

billion (as of 2011) and for wastewater infrastructure needs totals $298 billion (as of 2008).[1] Many rural communities face significant challenges in financing the costs of replacing or upgrading aging and obsolete drinking water and wastewater infrastructure. The demand for drinking water and wastewater infrastructure projects in communities with populations of 10,000 and fewer people, which are often considered rural communities, is estimated by federal agencies to be more than $140 billion in the coming decades.

A number of federal agencies provide grants, subsidized loans, and technical assistance to rural communities for drinking water and clean water infrastructure development. This testimony draws on our reports reviewing EPA's Drinking Water and Clean Water State Revolving Funds (SRF), the largest sources of federal funding for water and wastewater infrastructure, as well as programs managed by the Departments of Agriculture (USDA); Commerce; Housing and Urban Development, Health and Human Services; and the Interior; and the U.S. Army Corps of Engineers.[2] My statement today is based on our prior reports issued from September 2007 through October 2012, for which we updated information as appropriate through February 2015. My statement focuses on (1) the federal agencies that provide funding or technical assistance for rural communities and fiscal year 2014 funding and (2) issues identified in our work that affect rural communities' ability to fund water and wastewater infrastructure. Detailed information about scope and methodology for this work can be found in each of our issued reports. The work upon which this statement is based was conducted in accordance with generally accepted government auditing standards.

BACKGROUND

Across the country, about 52,000 community water systems provide drinking water to communities, while more than 16,000 wastewater plants treat sewage and return it to a nearby water body.[3] Like all communities across the nation, rural communities often have old or aging drinking water and wastewater systems. Federal agencies define rural differently, depending on agency guidelines and individual project or program authorizations. USDA's Rural Utilities Service provides funding to communities of 10,000 or fewer people. EPA considers communities with 10,000 and fewer people as small. According to EPA, small public water systems have difficulties meeting the requirements of the Safe Drinking Water Act to provide safe drinking water to their customers. The smallest water systems, those serving fewer than 3,300 persons, make up only 8 percent of the population served, but represent nearly 83 percent of all systems with reported funding needs, according to EPA.

The need for a water infrastructure project can arise for multiple reasons, including bringing a system into compliance with wastewater or drinking water standards and replacing or upgrading aging equipment. For example, arsenic is often present naturally in groundwater, and many rural communities that use groundwater may need to upgrade drinking water treatment systems to meet federal standards for arsenic. Others may need to update basic wastewater systems, which treat wastes by allowing them to settle out in ponds or lagoons, with more sophisticated equipment that mechanically and biologically removes solids and contaminants.

Our previous work on water infrastructure needs identified serious problems in providing safe drinking water and sanitation (wastewater) infrastructure in one largely rural region of the country, the region along the U.S.-Mexico border.[4] U.S. residents along the border who do not have safe drinking water supplies often buy and store water in outdoor tanks or barrels for drinking and other domestic uses. This practice represents a significant health risk because the water is often transferred or stored in open containers and is subject to contamination. In addition, some residents rely on substandard septic systems or cesspools to dispose of sewage because they do not have access to wastewater treatment systems.[5]

Seven Federal Agencies Provide Funding and Technical Assistance for Rural Drinking Water and Wastewater Infrastructure

Federal agencies administer a number of programs that provide funding and technical assistance to rural communities to help them build water and wastewater systems and comply with federal regulations. EPA's Drinking Water and Clean Water SRF programs and the USDA's Rural Utilities Service program provided most of the funding, but there are a total of seven key agencies that provide funding for rural water infrastructure:

- EPA annually provides grants to states to help finance local drinking water and wastewater projects nationwide through the Drinking Water and Clean Water SRF programs, some of which goes to rural communities.[6] States use this funding, along with a required 20 percent match, to capitalize their state revolving funds. The funds provide low-cost loans or other financial assistance for a wide range of water infrastructure projects. In addition, EPA provides funds from the Drinking Water and Clean Water SRF programs to tribal nations throughout the United States for water and wastewater projects.

- USDA provides grants, loans, and technical assistance for rural water and wastewater projects through its Rural Utilities Service.[7] USDA can provide assistance for various activities, such as construction of water treatment and sewage collection facilities, connection of single-family homes to water distribution or wastewater collection lines, and training for the operation of water and wastewater utilities. USDA targets its funding to communities with fewer than 10,000 inhabitants.
- The Department of Housing and Urban Development disburses grants to states and local governments through the Community Development Block Grant Program to fund housing, infrastructure, and other community development activities. The annual appropriation for the block grants is split according to formulas, so that 70 percent is allocated among eligible metropolitan cities and counties, and 30 percent among the states to serve cities with populations of fewer than 50,000 and counties with populations of fewer than 200,000.
- The Department of Health and Human Service's Indian Health Service constructs water and wastewater projects through its Sanitation Facilities Construction Program. This assistance is available to tribal nations within the United States, and through the program, Indian Health Service constructs various projects, including distribution and collection lines, treatment facilities, and home connections.
- The Department of Commerce's Economic Development Administration provides grants to economically distressed areas through its Public Works and Economic Development Program.[8] Grant funds are to be used for construction of public facilities, including water and wastewater facilities.
- The U.S. Army Corps of Engineers (Corps) provides assistance for water and wastewater projects in rural communities as directed by Congress. Congress has authorized and appropriated funds for the Corps to provide assistance for a number of projects, including projects that benefit rural communities in need of water or wastewater infrastructure.
- The Bureau of Reclamation (Reclamation) can be directed by Congress to provide assistance for drinking water or wastewater

treatment projects in response to individual project authorizations. In 2006, Reclamation received authorization, under the Rural Water Supply Act, to establish a rural water supply program.

Table 1. Selected Water and Wastewater Infrastructure Programs, Fiscal Year 2014 Funding and Fiscal Year 2015 Funding Request

Dollars in millions			
Agency and Program or Project	FY2014 funding	FY 2015 funding request	Type of financial assistance
EPA, Clean Water State Revolving Fund Loan Program	$1,449	$1,018	Grants to states to capitalize loan funds
EPA, Drinking Water State Revolving Fund Loan Program	$907	$757	Grants to states to capitalize loan funds
USDA, Rural Utilities Service, Water and Waste Disposal Program	$485	$247	Grants
Housing and Urban Development, Community Development Block Grant	$3,030	$2,800	Formula grants
Department of the Interior, Bureau of Reclamation, Rural Water Supply	$67	$34	Loans and grants
Economic Development Administration, Public Works and Economic Development Program	$96	$85	Grants
Indian Health Services, Facilities and Environmental Health Support	$211	$221	Facility construction and technical assistance
U.S. Army Corps of Engineers	Est. $45 (general) $38 (environmental infrastructure)	$26 (general)	Loans Technical planning services or grants

Source: GAO analysis of Congressional Research Service and Indian Health Service data. | GAO-15-450T.

Note: Funding refers to appropriations for fiscal year 2014 and budget request amounts for fiscal year 2015.

As we have reported, EPA's SRF programs are the largest sources of federal assistance to states and local communities for water infrastructure development. From fiscal years 2005 through 2014, funding for the Drinking Water SRF averaged $931 million annually and for the Clean Water SRF about $1.3 billion annually. In fiscal year 2014, the SRF programs were funded at $907 million and $1.45 billion respectively. See table 1 for fiscal year 2014 program funding by agency and requested fiscal year 2015 funding.

While these seven federal agencies can each provide funding for water and wastewater infrastructure in rural communities, they have varying eligibility criteria that may focus funding to specific communities on the basis of population size, economic need, or geographic location. For example the USDA Rural Utilities Service program specifically provides funding to communities with populations of 10,000 or less. The Drinking Water SRF program, however, prioritizes funding on the basis of environmental improvement but may offer special technical assistance to small communities, and must set-aside a portion of funding specifically for such communities. Similarly, the Economic Development Administration has established formal nationwide programs with standardized eligibility criteria and processes under which communities compete for funding. Specifically, the agency's criteria require projects to be located in economically distressed communities, regardless of the size of the population served, and the projects must save or create jobs. In contrast, Reclamation and the Corps have not historically had rural water supply and wastewater programs; rather, they have provided funding to specific projects in defined geographic locations under explicit congressional authorizations.

Issues Affecting Rural Communities Ability to Fund Water and Wastewater Infrastructure

Our October 2012 report on rural water infrastructure funding found several issues that affect rural communities' ability to obtain funding for water and wastewater infrastructure.[9] These issues include financing, technical expertise, and agency coordination.

Communities typically paid for drinking water and wastewater infrastructure through the rates charged to users of the drinking water and wastewater systems. In some cases, however, these communities did not have the number of users needed to share the cost of major infrastructure projects while maintaining affordable user rates. In addition, unlike larger, urban communities that can issue their own public bonds to pay for major water and wastewater infrastructure improvements, it can be difficult for rural communities to independently finance such major improvements. In many cases, rural communities had limited access to financial markets, restricting their ability to issue bonds to raise capital. As a result, these communities depended heavily on federal and state grants and subsidized loan programs to finance their water and wastewater infrastructure projects.

Rural communities did not generally have the technical expertise to rebuild or replace their water and wastewater systems. As we reported in October 2012, they had few staff and often hired consultants and engineers to help them design a water or wastewater project, including developing preliminary engineering plans and environmental documents. Generally, once a community determines that it will build an infrastructure project, it seeks funding from various sources. Once funding is confirmed, a community or its consultant designs the project, obtains construction bids, contracts to build the project, and, if applicable, seeks reimbursement from funding agencies. Some federal and state programs pay for technical service providers that communities can use to help communities design and finance their projects, and apply for funding.

Communities can also get assistance from local planning districts, which are voluntary associations of county and municipal governments that provide development assistance to their membership. In addition, according to EPA documents, in many cases, small systems may not be able or eligible to apply for loan assistance through the Drinking Water SRF because of a lack of technical, financial, or managerial capacity. EPA provides technical assistance through the Drinking Water SRF to help complete engineering studies to highlight operational areas for improvement, additional assistance in completing an environmental review and developing cost estimates during the planning process, or assistance completing an application for funding. The Rural Utilities Service also provides funding to nonprofits that provide technical assistance to communities with drinking water and wastewater issues. In 2011, EPA and USDA's Rural Utilities Service signed a memorandum committing to work together to help rural communities face the challenges of aging infrastructure, increased regulatory requirements, workforce shortages, increasing costs and declining rate bases.[10]

Our 2012 report also found coordination issues between the agencies that provide funding to rural communities for water and wastewater infrastructure.[11] Our analysis showed that EPA and USDA had not taken some of the actions specified in a 1997 memorandum of understanding on interagency coordination.[12] For example, we found potentially duplicative application requirements when applying to multiple state or federal programs, including preliminary engineering reports and environmental analyses, which may make it more costly and time-consuming for communities to complete the application process. Federal and state programs each have their own application process through which communities can apply for funding, although the application processes generally include similar steps. These steps are (1) completing an application that asks for, among other things, basic demographic, legal, and financial information associated with the project; (2) developing a preliminary engineering report that provides basic design specifications and other technical information for the project; and (3) conducting an environmental analysis that considers the environmental effects of the proposed project and alternatives. In some states we reviewed, communities told us that they had to complete separate but very similar application documents for the state SRF and state-level Rural Utilities Service programs. For example, in Colorado, an engineer for several projects we reviewed told us that the engineering firm had to complete preliminary engineering reports for both the state SRF programs and the state-level Rural Utilities Service program even though the reports had similar formats and information. In addition, in Colorado, North Carolina, and South Dakota, the state-level Rural Utilities Service program did not typically accept environmental analyses prepared for the SRF program because the state's analyses were less rigorous, according to Rural Utilities Service officials.

If consulting engineers have to provide similar, or even the same, information, in two different engineering reports or environmental analyses, their fees to the community may be higher. Engineers we interviewed estimated that preparing additional preliminary engineering work could cost anywhere from $5,000 to $50,000 and that the cost of an additional environmental analysis could add as little as $500 to a community's costs or as much as $15,000. Moreover, having to complete separate preliminary engineering reports or environmental analyses may delay a project because of the additional time required to complete and submit these documents. State officials in Montana told us that coordination between federal and state programs and the implementation of uniform application requirements could reduce the time it takes an applicant to complete a rural water infrastructure project by up to half.

In our December 2009 and October 2012 reports, we recommended several actions to improve coordination among the agencies and programs, such as completing an effort to develop guidelines to assist states in developing their own uniform preliminary engineering reports to meet federal and state requirements and starting an effort to create uniform guidelines for environmental analyses that could be used, to the extent appropriate, to meet state and federal requirements. As of February 2015, EPA and USDA had taken, or are in the process of taking, several steps to respond to our recommendations to improve interagency coordination on rural water infrastructure projects:

- EPA and USDA, working with other agencies including the Department of Health and Human Service's Indian Health Service and the Department of Housing and Urban Development, have developed a uniform preliminary engineering report template and associated guidance, according to EPA and USDA officials. The template has been adopted at the national level by EPA, USDA, and the Indian Health Service, and USDA began web-based training for field staff in May 2013. The agencies worked with 16 states to develop the template and, according to USDA officials, 7 adopted the template for their use, and other states may be using it informally.
- EPA and USDA have begun taking steps to develop guidelines to assist states in developing uniform environmental analyses. The agencies are participating in a work group, along with other federal agencies that fund rural water infrastructure projects (Department of Housing and Urban Development, the Department of Health and Human Services' Indian Health Service, and the Department of the Interior). The work group has developed a matrix that describes the environmental analyses required by each agency that is to be a tool that communities applying for funding can use to identify and complete appropriate environmental analyses. In addition, EPA is working with USDA to determine the extent to which requirements for EPA and USDA water infrastructure projects duplicate one another. As of February 2015, the agencies have drafted a best practices memorandum that highlights key issues found during this work, as well as what some states are doing to foster interagency collaboration, including reducing the potential for duplication of effort during the environment review process. According to officials, the memorandum is being reviewed within EPA and USDA and is to be sent to the states for their review and comment.
- The agencies are also taking steps in the U.S.-Mexico border region to coordinate policies and procedures and to prioritize funding for projects. For example, several agencies are participating in the Tribal Interagency Task Force (e.g., EPA, USDA, Department of Housing and Urban Development, Indian Health Service, and Department of the Interior), a group that coordinates drinking water and clean water infrastructure projects on tribal lands, including the U.S. Mexico

border. In addition, in 2014 EPA and USDA published a report describing a joint effort to address the critical public health and environmental challenges in the U.S.-Mexico Border region. This effort was created to leverage collective resources to assess needs within the border region. USDA announced a grant to fund the needs assessment in April 2014 and awarded the grant in June 2014, with a report expected in 2015.

In conclusion, the nation's water and wastewater infrastructure needs are large and funding them will be challenging. Rural communities face additional challenges in funding their water and wastewater infrastructure needs, given the financial, technical expertise, and coordination challenges they face overall. Federal agencies, with states, have long played a role in assisting local communities and will likely continue to do so. As they do so, they can and should consider how to ease communities' efforts to obtain funding. In addition, technical assistance and better coordination of federal efforts could help communities' address these challenges.

Chairman Shimkus, Ranking Member Tonko, and Members of the Subcommittee, this concludes my prepared statement. I would be pleased to answer any questions you may have at this time.

End Notes

[1] These data are from EPA's drinking water needs assessments. The most recent needs assessment is EPA, Drinking Water Infrastructure Needs Survey and Assessment: Fifth Report to Congress, EPA 816-R-13-006 (Washington, D.C.: April 2013) and the most recent clean water needs assessment is EPA, Clean Watersheds Needs Survey 2008: Report to Congress, EPA-832-R-10-002 (Washington, D.C.). EPA conducts a separate survey and assessment for each type of infrastructure, drinking water and wastewater, on separate 4-year schedules. The costs shown reflect the 20 year projected drinking water and wastewater infrastructure costs starting with the year that each survey was conducted.

[2] GAO, Water Resource: Four Federal Agencies Provide Funding for Rural Water Supply and Wastewater Projects, GAO-07-1094 (Washington, D.C.: Sept.7, 2007); GAO, Rural Water Infrastructure: Improved Coordination and Funding Processes Could Enhance Federal Efforts to Meet Needs in the U.S.-Mexico Border Region, GAO-10-126 (Washington, D.C.: Dec. 18, 2009); and GAO, Rural Water Infrastructure: Additional Coordination Can Help Avoid Potentially Duplicative Application Requirements, GAO-13-111 (Washington, D.C.: Oct.16, 2012).

[3] A community water system is a public water system that serves at least 15 connections used by year-round residents or that regularly serves at least 25 residents year-round. Cities, towns,

and small communities such as retirement homes are examples of community water systems.

[4] GAO-10-126.

[5] GAO-10-126.

[6] The Drinking Water SRF Program was established to make funds available to drinking water systems to finance infrastructure improvements. The Clean Water SRF Program was established to fund wastewater treatment projects

[7] USDA defines a rural area as a city of 10,000 or fewer residents or any unincorporated area.

[8] The Economic Development Administration defines an area as economically distressed if it meets one of the following three conditions: (1) an unemployment rate that is at least 1 percent greater than the national average, (2) a per capita income that is 80 percent of the national average or less, or (3) the area has experienced or is about to experience a special need arising from sudden and severe changes in economic conditions.

[9] GAO-13-111.

[10] Memorandum of Agreement between EPA and the Rural Utilities Service: Promoting Sustainable Rural Water and Wastewater Systems, 2011.

[11] GAO-13-111.

[12] The joint memorandum, signed in 1997 by EPA and the USDA, sought to improve coordination among federal and state agencies as they help fund community projects. It identified four major actions that state and state-level federal offices can take to improve coordination and reduce inefficiencies and potential duplication of effort, including: cooperation in preparing planning documents; cooperation in removing policy and regulatory barriers; cooperation on project funding; and cooperation in preparing environmental analyses and meeting other common federal requirements.

In: Rural Water Systems
Editor: Gloria Simmons
ISBN: 978-1-63483-240-3

Chapter 3

TESTIMONY OF ROBERT STEWART, EXECUTIVE DIRECTOR, RURAL COMMUNITY ASSISTANCE PARTNERSHIP. HEARING ON "THE NEEDS OF DRINKING WATER SYSTEMS IN RURAL AND SMALLER COMMUNITIES"*

SUMMARY

Small and rural drinking water systems constitute nearly 85% of the 53,000 community water systems in America. With limited staffing and ongoing issues related to accessing capital for improvements and repairs, these systems continue to have the highest violations of the Safe Drinking Water Act. Small water systems need increased access to USDA Rural Development Water and Environment Programs and for the EPA State Revolving Funds to be better managed to meet small system needs. Facing increasing regulatory demands, the need for planning and asset management programs along with requirements regarding proper financial management (and many others), these small utilities needs access to training and technical assistance. The most effective technical assistance programs place an experienced professional at the utility to provide guidance and support on technical, managerial, and financial requirements. Robust training programs are needed to educate operators,

* This is an edited, reformatted and augmented version of testimony presented February 27, 2015 before the House Committee on Energy and Commerce, Subcommittee on Environment and the Economy.

managers, and the governing body concerning their responsibilities and how best to operate and manage a small utility under a regulatory environment. Trainings need to be on-site, or within a localized area for easy access by the utility. In addition, more training should be delivered electronically to take advantage of reaching a larger number of systems without requiring expensive and disruptive travel requirements. Alternative service delivery approaches should be considered, such as sharing services, cooperative operations and management, and even full consolidation. Training tools and operational programs must be developed in order to be easily assimilated by small systems. And finally, new approaches to financing small utilities should be considered, noting that the new WIFIA program will not benefit small systems.

INTRODUCTION

Thank you, Chairman Shimkus, Ranking Member Tonko, and members of the subcommittee, for this opportunity to address the needs of drinking water systems in rural and smaller communities. In my nearly thirty years of work in the rural utility field, including twenty years in my home State of Texas, and now managing a nation-wide rural community development organization, I have experienced firsthand the many issues that impact the ability of small drinking water systems to meet regulatory requirements while providing their customers with safe and affordable services to promote the quality of life in the thousands of rural communities that form the backbone of our heartland.

My name is Robert Stewart, and I am the Executive Director of the Rural Community Assistance Partnership (RCAP). RCAP is a non-profit national network of regional service providers that for over 40 years has helped small, low-income, rural communities address water, wastewater, and other community development needs in all 50 states, Puerto Rico, and the Virgin Islands. Our team of community assistance providers delivers onsite training and technical assistance to small water and wastewater systems to help them meet regulatory requirements, finance and manage capital improvement projects, and to develop and sustain technical, managerial, and financial capacities.

For many years, the RCAP network has worked with USDA Rural Development and the Environmental Protection Agency as well as state primacy and infrastructure funding agencies to ensure that critical financing, technical assistance and comprehensive training opportunities are made available to small rural drinking water systems. RCAP not only assists rural communities with funding applications and every phase of the project planning and development process, but also provides training and technical assistance after construction is complete, helping communities understand how to properly manage and operate their system in a fiscally sustainable manner. Every year the RCAP network helps roughly 2,000 rural communities address their water and wastewater needs.

Providing these basic services is a challenge for many rural communities. Rural residents are three times more likely than their urban counterparts to lack water and sanitation; they also typically pay nearly three times the amount for water and sewer services. Due to their limited customer base, small utilities lack the economies of scale that reduce the costs of infrastructure construction, operation, and maintenance to levels that are affordable to low-income residents.

Access to Capital

As with their larger urban counterparts, drinking water systems in rural and smaller communities require access to capital to extend services and maintain treatment and distribution systems that serve their residential and commercial customers. There is no need to recount in detail here the extent of the need for water utility infrastructure investment over the next twenty years; these range from EPA's estimate of nearly $400 billion to those produced by the American Water Works Association and the American Society of Civil Engineers that place that figure closer to $1 trillion including over $59 billion just to meet the needs of drinking water systems serving under 3,300 population. While large utilities can issue municipal bonds for these improvements only about 4% of the 53,000 community water systems are large enough to issue their own bonds. For the 47,000 medium, small, and very small systems, the municipal bond market is not an option. These systems must depend upon federal and state financing sources such as those operated by the USDA Rural Utility Service's Water and Environmental Program (WEP), EPA's State Revolving Funds (SRF), other state specific infrastructure funding programs, or their own resources. Access to this capital is crucial to protect public health and allow for economic development in rural communities. Without this basic infrastructure, local employers will relocate or close factories and small businesses will decline and eventually disappear. The entrepreneurs and small business owners who are the engines of our economy won't open new businesses, shops, or restaurants on Main Street without basic services. Infrastructure remains a primary foundation of economic development, and to promote economic growth in rural America, businesses' and residents' basic needs, like water and sewer services, must be met. Opportunities for continued economic growth in rural communities are substantial. Agricultural production, oil and gas development, mining operations, alternative energy pursuits, and tourism are all vibrant economic sectors that depend on sustainable rural communities.

Small rural communities therefore need the continuation of EPA's and USDA's water infrastructure funding programs and with the extent of the need and the benefits to be derived, consideration should be made to increasing funding levels to these vital programs. EPA's SRF program should be better targeted at small rural communities. That program, created under the Safe Drinking Water Act Amendment of 1996, was to assist utilities in complying with increased regulatory requirements. Approximately 96% of all health based violations occur at systems serving a population of less than 10,000 while less than a third of the SRF outlays are directed at these same small systems. Small rural systems lack the economies of scale enjoyed by large urban systems resulting in higher user fees and a reduced ability to self-finance improvements. The average water bill in small rural communities funded by RUS is approaching $50/month (sometimes much higher for surface water systems), while at my home in Fairfax County I pay approximately that amount for water and sewer service combined. Even more important for rural communities has been the Rural Utility Service's WEP. The agency boasts a portfolio of more than 18,000 active water/sewer loans, more than 19 million rural residents served, and a delinquency rate of just 0.18%. This success is partly attributable to the field presence RD has historically maintained in rural areas. With staff in field offices throughout the country, RD is uniquely positioned to evaluate the creditworthiness of small utilities and is able to distribute federal funds quickly and efficiently to areas of great need. Staff reductions in RD offices across every state have started to hinder the ability of RD to serve rural communities with critical services. In drought years, or after natural disasters, community leaders benefit from being able to turn to a local RD staffer that they know and trust and who is familiar with their system and its needs. RCAP supports the continuation and strengthening of the WEP as a primary means to meet drinking water needs in rural communities.

Technical Assistance and Training

Small water utilities need increased access to technical assistance resources and training programs that can enable them to evaluate and obtain capital financing, operate in accordance with regulatory requirements, and cost-effectively manage their utility. Small city council or water district directors are most often volunteers who lack professional staff and the resources to find out what funding sources are available or the requirements for funding eligibility. Funding application processes and eligibility requirements for each federal and state program are slightly different and each poses unique challenges. With help from an experienced technical assistance provider, however, even communities with no staff and limited planning resources can develop the local leadership capacity to apply for and manage needed infrastructure projects. Technical assistance plays a vital role in ensuring that the programs serve the communities they were designed to benefit in a cost effective manner. While there are many calls for reducing the requirements associated with obtaining water and wastewater financing from RD and EPA, RCAP's opinion is that these requirements are for the most part necessary to ensure that the federal government is making financial support available to the neediest communities while ensuring the security of the federal investment. The extremely low default rate on these loans is a testament to the efficacy of existing requirements. Common environmental review requirements among all federal and state infrastructure programs would be one area for improvement. Oftentimes projects have multiple funding sources with varying environmental review/assessment requirements causing duplication of effort that produces no tangible benefit to the funders.

Technical assistance and targeted training is critically needed to assist small communities meet increasing regulatory requirements. Small and very small systems (depending on their size) typically employ a single certified water operator who is responsible for all operations, maintenance and repair of the utility's treatment plant and distribution system. These professionals are typically overworked and underpaid and the majority are nearing retirement age. My experience has been that these operators are extremely dedicated to their work and will do everything possible to ensure that their customers receive uninterrupted water service that meets all requirements. However, new federal rules such as the Revised Total Coliform Rule place addition demands on operators who must learn about this rule and apply it to their utility. On-site technical assistance provides these operators with the guidance and support they need to meet operational requirements and ensure public health. In almost all cases the only way this assistance is provided is through on-site assistance provided by RCAP's Technical Assistance Providers (TAPS) or by state Rural Water Associations' Circuit Riders. On-site assistance is also needed to help the utility manage their business and financial affairs. Preparing budgets, conducting rate studies, developing Operation and Maintenance manuals, preparing customer service policies, and many other similar requirements are extremely difficult for these small utilities. TAPS and circuit riders provide the tools, the expertise, and the guidance to help small utilities with these programs. While both EPA and USDA Rural Development acknowledge the efficacy of on-site technical assistance, funding for these programs has decreased in recent years. Technical assistance funding authorized under Section 1442(a) of the Safe Drinking Water Act ($15 million a year) is never included in the administration's budget, requiring Congress to add an amount ($12.7 million in recent years) in their appropriation. This amount authorized nearly 20 years ago is not sufficient to meet the needs of small water utilities and should be increased. In addition, Section 306(a) of Consolidated Farm and Rural Development Act allows USDA to fund technical assistance grants and a circuit rider program (recently at $19 million and $15 million a year respectively). An increase in funding levels to these programs would help better meet the needs of rural water systems.

In order to meet the varied needs of small communities, RCAP in recent years has created a multi-faceted program to deliver a variety of training resources. Oftentimes small utilities cannot afford to have their only operator travel for centralized training classes. As such, much of RCAP's training is conducted on-site, whether it is directed at the operator, the manager, or the governing body of the utility. In addition, RCAP has designed and is presenting a variety of synchronous and asynchronous web-based trainings that allow utility staff and managers to access training from their workplace or home. RCAP places a wealth of information on utility operations, management and financing on our website, available free of charge. These informational materials are prepared specifically to meet the needs of small utilities, are easy to understand and apply to local requirements. RCAP strongly supports the provision of training services through on-site training, local or regional classroom style training, on-line training and the provision of educational materials both in-print and online. Only through this multi-facet approach is there any expectation that small utilities will be able to access the training resources that they need.

Sharing of Services, Cooperative Approaches and Regionalization

Another means to assist rural utilities is to provide guidance, support, assistance, and incentives in the areas of sharing services or using cooperative approaches to deliver services. These can include everything possible up to and including consolidation or regionalization of services. In order to maximize limited resources, communities need to realistically examine whether operating their own facilities is cost effective. Services such as meter reading, billing, purchasing, or employing a certified operator can be shared among two or more nearby utilities. With respect to water infrastructure, at times clusters of small towns can better and more affordably be served by having one large treatment plant with pipes running to each town than by having a separate treatment facilities. In areas where communities are too far apart to run pipes, utilities could benefit from shared management, operations, purchasing, and other similar joint service provision. Regionalization may not be feasible in all cases, especially in areas with large distances between communities. However, RCAP recommends that potential borrowers demonstrate to RD and the SRFs their efforts to employ regionalized service provision as part of the application process. Most states now require that new or expanding utilities provide documentation regarding their efforts to regionalize prior to their being granted a license or certificate to serve an area. Priority should be given to applications for regional service provision, especially in cases where smaller or non-compliant systems are being consolidated. By giving priority to projects in which the applicants can demonstrate that they have weighed the costs and benefits of regionalization, RD and EPA can encourage regional projects where appropriate without disqualifying communities that are geographically isolated. RCAP's experience has been that regionalization is most often successful when a technical assistance provider is able to spend time with all entities involved to offer alternative approaches, assist in the evaluation of costs and benefits, identify funding sources, prepare necessary documentation, and assist with public education and outreach.

Tools to Improve Operations and Management

One strong point for EPA in particular has been its development of a variety of tools that can be used by small utilities, such as the Simple Tools for

Effective Performance (STEP) guides or the Check-Up Program for Small Systems (CUPSS), which is an asset management program (there are many others). In addition, EPA and RD have been working together to produce a variety of tools that address effective utility management and planning. This is one role that the two agencies can effectively assume to help small systems. RCAP would encourage the agencies to continue to seek out and incorporate advice from RCAP and the National Rural Water Association. Both of our organizations have the means and experience to ensure that these tools meet the needs of the small communities and the ability to deliver these tools on-site in rural communities. Other tools and programs needed by small communities that have been developed or used by RCAP include, ones that address vulnerability assessments, emergency response plans, contingency plans for drought and storm-related events, use of GIS systems to map utility components, energy audit programs, and leak detection programs. Most of these tools have widespread applicability so that the initial investment in the tool, process, or program can be repaid many times over through multiple uses in rural communities across America.

Operator Training

This is a specific training issue that I wanted to briefly address as it impacts all small systems. All water operators must receive training and pass certification exams in order to operate public water systems. Continuing training is required in order to keep operator licenses current. For small systems, paying for this training and allowing the operator to leave the utility site is problematic. As a result of the 1996 Amendments to the SDWA the Operator Certification Expense Reimbursement Grant Program was created to fund training needs for systems of fewer than 3,300 populations. Funding for this program expired several years ago. RCAP would recommend that consideration be made to reauthorizing a similar program that would assist small systems in operator certification requirements. Any renewal of this program should emphasize on-site, hands-on and experiential training that is most needed and most effective for small water system operators.

The Water Infrastructure Finance and Innovation Act (WIFIA) and Other Options WIFIA

WIFIA was included as part of last year's Water Resources Development Act (WRDA), as a means to provide an alternative funding source for water infrastructure. RCAP is concerned that WIFIA could have a detrimental impact on the State Revolving Funds (SRFs) if states are allowed to divert SRF funds to meet WIFIA's 49% non-federal matching requirement. WIFIA must not impair the ability of the SRFs to provide critical support that our nation's rural water and wastewater systems need. As mentioned earlier, due to their small customer bases and difficulty in accessing the bond market, these rural systems rely on the availability of SRF funds to fund the upgrades necessary to comply with the requirements of the Safe Drinking Water Act and Clean Water Act. It is therefore imperative that WIFIA be implemented in a manner that does not infringe upon the SRFs, and that ensures that rural communities are able to meet their obligations to provide their residents with safe drinking water and sanitary wastewater services.

WIFIA's authorization language is unambiguous: Congress clearly intended to have WIFIA serve as an additional tool for water infrastructure financing, and not as a replacement for the SRFs. The Senate Environment and Public Works Committee's report states that WIFIA "offers the sponsors of these water infrastructure projects a new tool to...stimulate additional investment in our Nation's water resources infrastructure" (Senate Report 113-13, emphasis added). Further, WRRDA §5028(a)(6) requires EPA to notify the applicable SRF when a WIFIA application is received, and gives the SRF the right of first refusal. Together, these provisions clearly demonstrate that Congress intended for WIFIA to supplement, not to replace, the SRFs. Even WIFIA's proponents never advocated for WIFIA to be a substitute for the SRFs. Mr. Matthew Millea, testifying before the Interior, Environment, and Related Agencies Subcommittee of the House Appropriations Committee on behalf of the Water Environment Federation said "WIFIA must be designed to complement—not replace—the SRFs." Mr. David Weihrauch, testifying before the Water Resources and Environment Subcommittee of the House Transportation and Infrastructure Committee on behalf of the American Water Works Association said "WIFIA is designed to supplement the SRF by addressing needs that are not well addressed, if at all, by the SRFs." Taking money out of the SRFs to pay for WIFIA projects would therefore contradict the clear intent of the authorizing legislation.

Among the statutory objectives of the Safe Drinking Water Act provisions creating the Drinking Water SRFs is that the funds "assist systems most in need on a per household basis according to State affordability criteria." The systems that are most in need on a per household basis are most often small, rural systems. Many factors contribute to this reality, including persistent rural poverty, limited water resource availability, lack of population density (requiring longer segments of pipe between connections), and lack of economies of scale. In addition, most rural systems, unlike their urban and suburban counterparts, are unable to obtain credit ratings necessary to access the municipal bond market, thereby eliminating one potential source of affordable financing. As noted earlier, a recent Johnson Foundation study estimated that only 2.8- 3.8% of water systems are large enough to issue their own bonds. These factors and others create severe affordability challenges in rural areas that the SRFs help to address. The WIFIA program, by its very nature, exists to serve large-scale projects that can attract private investment. In order to attract private capital, the projects are likely to serve more affluent areas and areas with large numbers of ratepayers because those areas are considered to be less of a credit risk. By definition, however, those areas are also those that have the fewest affordability concerns (more ability to repay equals less credit risk). Using SRF funds for WIFIA projects would therefore subvert the intent of the SRFs by taking funding from those communities that most need it to provide affordable services and diverting it to those who least need the assistance. This is consistent with neither the statutory objectives of the organic statutes that created the SRFs nor the clear Congressional intent to have WIFIA serve as an additional tool to attract new investment to water and wastewater infrastructure.

As such, WIFIA should be implemented in a way that compliments the SRFs by providing an additional funding mechanism for those communities that do not have affordability issues and are able to attract private investment, as envisioned by the law's proponents. The focus of WIFIA should be to attract additional investment to our nation's water and wastewater infrastructure, not reallocate the limited existing funds that help make compliance with federal and state regulations affordable for many smaller systems. The SRFs are effective programs that have a proven record of success. They have provided billions of dollars in affordable financing for communities that could not otherwise afford the costs of compliance with water and wastewater regulations. They are essential tools in the protection of public health and the environment in rural areas and should not be undermined by the implementation of WIFIA.

Drinking Water SRF

Application requirements should be simplified for small communities seeking to be listed on the state's Intended Use Plan in order to minimize costs to the community and to allow them to be considered for funding. States could require just a simple statement of water deficiencies, a reasonable proposal to correct them and a practical estimate of costs. Along with data already collected by the state, especially regarding water quality issues, the state should be able to properly list in a prioritized manner those small systems eligible for SRF funding. More detailed information could be obtained after a community has been approved for the intended use plan. In addition, once a small community project is set for inclusion on the Intended Use Plan, the state should provide small communities with technical assistance to work through the complete application process. USDA offers technical assistance for this purpose resulting in an increased ability for small systems to meet application requirements in a timely manner.

The SDWA permits states to set aside 15 percent of the DWSRF to finance projects in small communities. As has been reported the unmet needs of small communities are great and financing unavailable. Congress should consider a requirement that 25% (or more) of the DWSRF must be set aside for use by small communities.

A portion of the DWSRF should be dedicated to providing capitalization grants to qualified nationwide nonprofit intermediaries to establish revolving loan funds to help small communities finance pre-development costs and system repairs up to $150,000. While small communities may receive reimbursement for pre-development costs once SRF finances a project, they often do not have the resources to undertake these activities. A program that provides assistance for activities such as preliminary engineering studies or site acquisition would reduce this barrier to funding. A similar financing gap exists for small repairs needed by rural water systems. As a result of high transaction costs most state SRF programs do not accept (or will not end up funding) applications for projects under $200,000. Small utilities typically do not have the credit history to access these funds from local financial institutions. Currently, a USDA program allows for a small amount of their funds to be used to capitalize such revolving loan funds, currently $1 million a year. Both RCAP and NRWA operate their own revolving loan funds, but need additional capitalization to meet the tremendous need from small communities for this type of alternative funding arrangement.

Common Application Requirements

While some states have established uniform application requirements more work is needed in order to standardize applications for federal funding assistance. USDA and EPA have worked together to establish common requirements for preliminary engineering reports. This was an important first step, however, environmental reviews and other federal requirement should be common to all drinking water applications so that small systems are not required to understand and comply with different requirements from federal funders.

Final Thought

There has been a considerable amount of time and effort expended to develop and enact the WIFIA program. While this program can be effective in providing additional financial resources to large water systems, its ability to do so for small systems is highly problematic. What is needed is an approach that targets the capital and financing needs of small communities. This could come from changes made to existing programs, additional funding for these existing programs or even consideration of new programs such as a Water Trust Fund similar to that in place for transportation. What is certain is that small utilities are undercapitalized, struggle to meet the costs of new regulations, and suffer from diseconomies of scale. Small water systems pride themselves on being self-sufficient, a reflection of American values still dominant in rural America. However, as summarized in this discussion, there are many approaches that can be taken to provide support and assistance to these small water utilities; community based, community operated, managed and governed organizations that are an indispensable foundation to rural life and the rural economy so crucial for America.

In: Rural Water Systems
Editor: Gloria Simmons
ISBN: 978-1-63483-240-3

Chapter 4

TESTIMONY OF JOE KEEGAN, MAYOR, CASTLETON ON HUDSON, NEW YORK. HEARING ON "THE NEEDS OF DRINKING WATER SYSTEMS IN RURAL AND SMALLER COMMUNITIES"*

Good morning Mr. Chairman and Members of the Subcommittee, and my Congressman, good morning Congressman Tonko.

I am Joe Keegan, the mayor of a very charming little village a few miles south of Albany on the banks of the Hudson River called the Village of Castleton. We have a population of approximately 1,500 of the best people anywhere.

My village is a member of New York Rural Water Association, a non-profit organization of small and rural communities throughout the state which is somewhat responsible for my appearance here today. I got a call from the association on Monday asking about my availability and I just happened to be traveling back to Castleton last night from a trip related to my day job. Congressman Tonko knows this, but our part of New York is buried in snow and cold, and the other little village that was considering appearing, the Village of Delanson, said they could not come because the water department also

* This is an edited, reformatted and augmented version of testimony presented February 27, 2015 before the House Committee on Energy and Commerce, Subcommittee on Environment and the Economy.

has to plow the snow and they said they couldn't leave the plows even for Congress this week.

My village is very typical and representative of communities that have water supplies in New York and the rest of the country. According to the U.S. Environmental Protection Agency, the state of New York has 2,305 community water systems, 2,203 serve populations under 10,000, and, 2039 (88%) serve populations under 3,300. Correspondingly, the U.S. has 50,581 community water systems, 46,313 serve populations under 10,000, and 41,393 (82%) serve populations under 3,300. All of us small community water and sewer utilities have to comply with the same regulations, testing, and certifications as the biggest cities – but with only our very small rate-payer base. And we have to operate, maintain, and update our water infrastructure with very small budgets.

As a small community mayor, my number one concern and worry is drinking water and number two is wastewater – and everything else is a distance third. If there is a problem with the drinking water, it has to be addressed immediately, middle of the night, middle of the winter, it doesn't matter when. Every citizen and especially the most vulnerable depend on the safety of the water, including families with infants, schools, our nursing homes, and people with compromised immune systems. We can't have any potential contamination of the drinking water. Similarly, our sewer needs to function properly to avoid any possibility of a sewage spill or sewage back-up into people's homes.

I would say to you that this really does keep me up at night. Just last week, the frost penetrated the ground so deeply that we experienced two ruptures in our water mains that are 5-6 feet underground. This forced us to issue a boil water advisory where we have tell families they have to boil the water, as well as contact all the schools, the press, the nursing home, etc. – I actually call as many citizens as I can by robo-call. When something like this occurs, we are doing nothing but managing the situation around the clock and trying to resolve the situation, find the equipment to excavate the frozen ground, repair the water line, get the tests to the lab, and wait for all-clear results to lift the boil order. Every additional hour of having to boil their water is a serious hardship on a family. School drinking fountains have to be covered. Pallets of bottle of water have to be procured and delivered. Some acutely impacted homes have to move to hotels, friends, or family. My village and every small community wants to take every measure possible to avoid this situation and protect the public – our citizens.

We appreciate the assistance of this Subcommittee and the Congress in helping us protect the public and successfully operate the public drinking water and wastewater supply through the various funding programs and the on-site technical assistance initiatives. My village relies on this assistance.

I want to thank Congressman Tonko for sponsoring the "Assistance, Quality, and Affordability Act of 2014," in the last Congress. Small and rural communities support your legislation because it enhanced the current drinking water state revolving fund by further targeting the funding to communities most in need, included grants and not just loans for small communities who can't afford to repay loans (like mine), expanded the technical assistance initiatives that small communities depend on, and attempted to simplify the process to small communities so they can better understand the process and secure funding.

We do need help; most everything from financing, regulations, compliance, and the various programs is very complicated for small communities. We don't have financial professionals on staff and often don't understand many of the funding processes. Consequently, small communities often fail to exploit some of the funding programs like the state resolving funds. Currently, my community does not have any financing through these funds. We desperately need funding, and I could not tell you exactly why we don't have access to these funds.

We currently have needs approaching 3 million dollars for our wastewater system. We need new aeration tanks, new sludge drying equipment, and new pumps as our facility is 30 years-old. We need to stop rain water from leaking in the system and overtaxing our capacity. We need to update the process of injecting air into our aeration tanks, and replace the old roof on the plant. My water operator is constantly explaining to me the need for these upgrades and his concerns of possible failure. However, we really don't have a way to finance it. It would triple the sewer rates to take out a loan for that much. We just financed a new storage tank for the drinking water for 800,000 dollars using municipal bonding and we'll have to pay that off. You can see in the picture at the back of my testimony that we have some very old drinking water pipes that need updating or replacing at a substantial cost. The one in the picture is stamped with a date from the 19th century – and they are still in the ground in parts of the village. It is said these pipes came over from France as the ballast of ships and we purchased them for the water. We are concerned that without more water line replacement we are vulnerable to more breaks and crises. And you can see the other picture of a tuberculated pipe we recently dug up that is loaded with corrosion and deposits to the point it is almost occluded.

We need to replace these water mains. We recently had to repair a road over one of these mains, which I desperately wanted to replace – and the road work presented a convenient time to excavate and replace the main. However, the reality of our situation was the main replacement would have cost 2 million dollars. We couldn’t afford it, and we used all of state road allocation to resurface the road which means we can’t disturb the road for 10 years.

In my remaining time, I want to emphasize the essential assistance we receive from New York Rural Water Association technical assistance and explain why it is so helpful. The association has circuit riders that are on call throughout the state that will come and assist us immediately including evenings and weekends. The circuit riders are all experts in the technical side of water operations. Just a week ago, we called for help for locating a water leak from a ruptured pipe that could have occurred over any part of the 150 foot water line. The circuit rider has specialized equipment that can detect noises and vibrations underground to locate the exact location of a break – where we need to excavate. This saves our village significantly in road disturbance. We can’t afford to own the equipment and the circuit rider is an expert in using it – this saves us thousands of dollars and time. We have had their source water protection circuit rider assist us with compliance with the EPA mandate to delineate our source water area which saved us thousands in consultant fees. The wastewater circuit rider has analyzed the necessary repairs to our wastewater system and helped us prioritize the upgrades. My operators receive 90% of the training needed to retain their operators' licenses from New York Rural Water Association. My operators are so appreciative of the assistance funded through the Congressional appropriations process because the rural water circuit riders are always there and always available. We depend on them just like every other small community.

Mr. Chairman, I have a lot more to say, but you have been very charitable with your time and attention to small and rural communities – and on behalf of every small town elected official, we are grateful. Thank you from hearing from us and I will be happy to answer any questions.

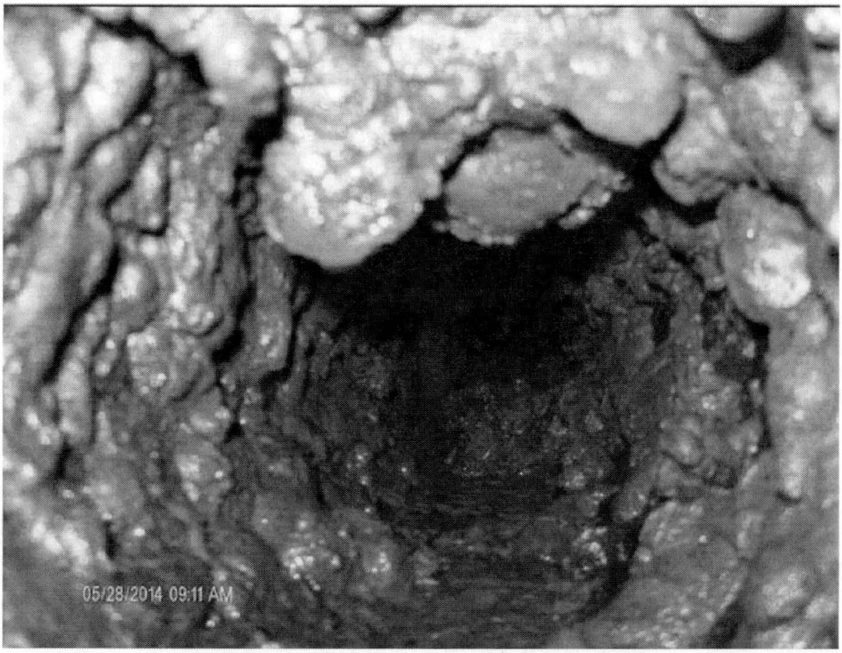

Water break, Stimpson and Second Streets

#10 Seaman Avenue, 8 inch main, replaced

In: Rural Water Systems
Editor: Gloria Simmons
ISBN: 978-1-63483-240-3

Chapter 5

TESTIMONY OF K.T. NEWMAN, ON BEHALF OF NATURAL RURAL WATER ASSOCIATION. HEARING ON "THE NEEDS OF DRINKING WATER SYSTEMS IN RURAL AND SMALLER COMMUNITIES"*

Good morning Mr. Chairman and Members of the Subcommittee, thank you for the opportunity to testify here today.

Before I introduce myself, I would like to summarize the key points I plan to make in my remarks:

- That most all the communities needing to comply with the federal EPA water mandates are small and rural communities.
- That these communities have a more difficult time complying with federal mandates and operating and maintaining their water infrastructure because of limited economies of scale, limited budgets, and limited technical resources.
- That when small and rural communities look for help for compliance, finding funding, operations, and emergency response – they all call their state rural water association on-site technical assistance providers – every time.

---

* This is an edited, reformatted and augmented version of testimony presented February 27, 2015 before the House Committee on Energy and Commerce, Subcommittee on Environment and the Economy.

- That of all the billions of dollars that Congress provides to the federal agencies for environmental programs, the only assistance that small and rural communities see and utilize is the assistance from these on-site "circuit riders."

My name is K.T. Newman and I have been working for or in small and rural community water systems in the Mississippi Delta for nearly 20 years. I first started out as a small city water manager in my hometown of Town of Vaiden, Mississippi which has about 1,000 homes. I then worked for the Mississippi Rural Water Association as a circuit rider for ten years. In this capacity I visited every one of the Delta's approximately 500 small communities to help them with their water and sewer problems. Currently, I am working for about two dozen small Delta communities, assisting them with their water and sewer utilities.

I am honored to be accompanied here today by the mayor of one of these small communities, Mayor Everette Hill from Como, Mississippi. The Town of Como has a population of approximately 1,200 persons. Mayor Hill has been mayor for two years and his community is facing overwhelming water challenges. Como is typical of the types of challenges many of the approximately 45,000 small and rural communities across the country are facing today regarding their water infrastructure. The mayor's challenges are compounded by the fact he is a small town mayor, meaning he has a full-time job (as a truck driver), has to handle much of the city's issues on his free-time, his community has little professional staff because they simply can't afford it.

In Como, the wastewater system is failing because of its age and inability to meet its current EPA permit. The cost to update Como's sewer system to be compliant is approximately 2 million dollars. The Como drinking water system needs an additional 1.0 million dollars in upgrades. The town was recently fined by the department of environmental quality for failure to comply with their wastewater discharge permit; currently the Como wastewater treatment facility is actually discharging only partially treated wastewater due to failure of the current treatment works. Within the past few months, Como finished paying the approximately 1 million dollar loan to construct their currently failing activated sludge treatment system. The loan placed considerable hardship on the residents.

Como is just like thousands of other small communities in the Delta and the other states, they need a grant-rich infrastructure funding program like the USDA's rural development program, and they need access to someone they can trust for technical advice, on-site assistance, and help with managing the funding application process.

Mississippi has 1,234 regulated public water systems -- only 2 serve populations over 50,000 persons and only 59 serve populations over 10,000 persons. That leaves 1,175 small and rural communities with populations under 10,000 persons that have to comply with every EPA regulations just like the larger cites of Jackson or Gulfport.

In the Town of Utica with a population of 850 persons, we are facing a nearly 1 million dollar compliance upgrade to meet our new and more stringent wastewater discharge permit. The town will likely have to accept hundreds of thousands of low-interest loan.– I can personally see the repayment of this loan placing significant hardship on the rate-payers. The Town is accepting this tremendous burden to pay for a new treatment technology called bio-domes that will be designed to enhance their current facultative lagoon cells and reduce the nutrient levels in the wastewater effluent. Small towns all across Mississippi and in fact the Country are faced with this dilemma. I believe small towns should be given more flexibility in their approaches to addressing these dilemmas. In addition, more training needs to be provided to small Town Mayors like Mayor Hill so that multi-million dollar upgrades that will most certainly tax the rate payers of these communities can be more readily understood and communicated to these residents who will ultimately be responsible for bearing the financial burden.

Recently, many of the small communities in the Delta have received violations for a new EPA regulation referred to as the "disinfection byproducts rule." These byproducts are a result of disinfecting their water to make it safe to drink. If these small communities limit or reduce the disinfectant levels of the water, they will comply with the EPA regulation, but the water may no longer be safe to drink. Once the "disinfection by products rule" is violated, many small communities are forced to spend limited resources to report these violations to the consumers As I am sure you can imagine, ay notice indicating a problem with the drinking water has the potential to cause pandemonium in these small communities. We urge Congress to take a look at this current and acute situation.

In the Town of Shaw, population 1,900 persons, the community was under a boil water order for over 6 months because of a broken chlorinator needed to disinfect the drinking water. The local schools had to buy bottled water for over six months. After they called the Mississippi Rural Water Association Circuit Rider, Tom Abernathy, they were able to come up with a plan to pay for a new chlorinator, revise the town's billing program to accurately assess the water used by citizens and receive the payments, train the new mayor and town council, get the Towns' credit stable, and secure some emergency state revolving fund financing. I have to acknowledge that Congressman Thompson was essential in getting this plan started for the initial assistance to the community. The new mayor is overwhelmed with the challenges the town is facing, but now he has a plan and someone he can count on and trust to help the town deal with all the challenges – and that person is the rural water circuit rider.

In closing, whenever a small community is facing a compliance issue, the complication of a new EPA rule, a line break that they can't find that is causing people to lose water service, an emergency from a storm or power loss – we all call the circuit riders to tell us what it means and what to do. They have developed a trust relationship with small communities in their states, know how to fix things, and are willing to come to your town day, night or weekend. The analyst here in DC who works for rural water told me that Congress gives EPA over 2 and 1/2 billion dollars every year to pay for environmental programs – I am sure that it goes to good use somewhere, but the only benefit that small and rural communities get to help them with their water issues comes from the on-site technical assistance provided by their state rural water associations. Thank you for the opportunity to be here, and Mayor Hill and I are happy to answer any questions.

In: Rural Water Systems
Editor: Gloria Simmons
ISBN: 978-1-63483-240-3

Chapter 6

TESTIMONY OF BOBBY SELMAN, ON BEHALF OF MISSISSIPPI RURAL WATER ASSOCIATION. HEARING ON "THE NEEDS OF DRINKING WATER SYSTEMS IN RURAL AND SMALLER COMMUNITIES"*

Good morning Mr. Chairman and Members of the Subcommittee, it is an honor to appear before you today.

My name is Bobby Selman. I am a certified drinking water and wastewater operator in the state of Mississippi with an engineering background from Mississippi State. I have been working in the water world for 25 years, starting in my home town in Lawrence County. I still work for the Lawrence County Water Authority in addition to 12 other small communities and rural water associations.

Before I give you some thoughts on the water challenges small and rural communities are facing related to more stringent compliance, depressed economies, aging of water infrastructure, failing infrastructure, limited technical expertise, new standards for operator certification, hurricanes devastating the water supply, and shrinking funding programs – I want thank my Congressman, Gregg Harper for his support and assistance to all the over 150,000 small public water systems across the country for sponsoring the

* This is an edited, reformatted and augmented version of testimony presented February 27, 2015 before the House Committee on Energy and Commerce, Subcommittee on Environment and the Economy.

Grassroots Rural and Small Grassroots Rural and Small Community Water Systems Assistance Act. Representative Harper's bill directs the U.S. Environmental Protection Agency (EPA) to prioritize the type of technical assistance that small communities find is most beneficial. The rural water type of on-site technical assistance is what all the small communities in Mississippi and the other states rely on for help with compliance, operations, emergencies, line break, loss of water, setting rates, and training for operator certification. I am told that Congress funds the EPA's internal management budget for hundreds of millions of dollars every year. Small and rural communities want Congress to know that the only benefit we get comes from the small portion of the EPA funding that is directed to on-site technical assistance provided by what we call circuit riders. Guidance documents, inspections, manuals, symposia, reports, on-line tools, webinars, and civil penalties don't really help at the local level when a small town needs immediate help. What small communities do when they have a question or water issue is call their local circuit rider that they know, trust, and know can give them clear answers – these circuit riders often come immediately on-site to small communities and teach them how to fix their problem. There is just no one else out in the field at the local level providing this essential help.

After Katrina, two of my small communities in Simpson County were devastated. Each served approximately 2,500 people and they were without power and water. People and communities can get by without power for a while, but not without water. I called the Mississippi Rural Water Association circuit riders and they found emergency generators for me and delivered them to the communities at no charge. Since the circuit riders know everybody in the state, they were able to borrow some generators from northern communities not impacted by the hurricane and have the generators delivered to get the drinking water and sanitation restored expeditiously. The circuit riders also have the technical know-how to rig the generators' electrical systems, size the right voltage, and even drive a back-hoe if needed to clear the streets and dig up rupture lines – all of this type of assistance is essential to restore a water supply in an emergency.

I called the circuit rider out to help me at the Double Ponds Water Association, a community of about 1,000 homes, to find a line break causing a loss of water for many homes. The circuit rider came with advanced radar equipment that can precisely identify the location of the break – which on this day happened to be out in the woods. By funding the circuit riders, Congress is allowing all small and rural communities to share this technical resource that no one community can afford on their own. We think it is the best use of your federal water-environmental dollars.

With the federalization of operator certification under the Safe Drinking Water Act of 1996, state rural water associations have become the main source of training for operators and the main source of continued education credits which are needed every year to maintain the certification. It can be very expensive for a small community to keep their operators certified, in fact many small communities can't afford to have a certified operator and have to share one – as is the case with many of the communities I am representing. Rural water training is often provided free of charge, provided locally, and tailored to exactly what that particular operator needs to stay certified and competently operate their specific facility. Many small community water systems only have a few hundred homes that have to fund their operations, infrastructure, testing, treatment chemicals, and personnel for both a water and sewer supply. Further, many of these systems were built in the 1960s and are reaching the end of their lifespans resulting in pipe breaks, tanks and pumps needing replacing, old wells failing, and the list goes on.

Many parts of rural America have seen industry move on, leaving behind depressed economies. In my region, the garment industry moved south after NAFTA. When this happens, raising rates becomes overly burdensome. In the town of New Hebron, Mississippi with just over 400 people, we are being told we need to comply with a new EPA wastewater discharge permit for the cost of 2-3 million dollars. In addition to more stringent compliance standards, standard operation and maintenance that a large community would have no problem with absorbing the cost can be very expensive on a cost per household basis in small water supplies. In Lawrence County with approximately 2,000 persons, we are trying to finance a new well because our 50 year-old well has rusted out and at the same time pay for the storage tank to be painted for a combined cost of 500,000 dollars – and this is a relatively large, small water supply.

It would be nice if we could consolidate all the little towns into one large water supply to realize cost savings, limit duplication, and obtain greater economies of scale. When this is practical, we emphatically pursue it. Often it is more expensive to develop the new infrastructure needed to accomplish regionalization than to continue to operate separate systems.

I will close with some comments on the federal water infrastructure programs – namely the EPA state revolving funds and the USDA rural development grant and loan program. We are very appreciative for Congressional funding of these initiatives, and realize the funding constraints on Congress and the nation. Notwithstanding the curtailment of federal funding, the regulatory burden continues to increase and become more complex. We urge you to emphasize grants in these funding programs. Low interest loans often don't help the communities facing the most severe hardship from federal compliance – leaving the loan funds to be used for compliance with greater ability to afford financing.

We also urge to keep reviewing the federal funding program to ensure that the subsidized funding is being targeted to the communities most in need financially and environmentally. And please do what you can to make the process of obtaining funding as simple as possible. Often the complexity of the funding process overwhelms small and rural communities' abilities – this could be addressed by using some of the infrastructure funding to help these communities with the application process. It can currently take up to 3-4 years from the beginning of the process to the awarding of funding.

We are very grateful for the funding assistance, it has allowed many rural and small communities to have access to drinking water and sanitation that they would otherwise not have been able to afford without the federal assistance, and we want to be partners in the effort to make the initiatives as efficient and successful as possible.

Thank you very much Mr. Chairman and I am eager to answer any questions at the appropriate time.

In: Rural Water Systems
Editor: Gloria Simmons
ISBN: 978-1-63483-240-3

Chapter 7

Rural Water Supply and Sewer Systems: Background Information*

Claudia Copeland

Summary

The Safe Drinking Water Act and the Clean Water Act impose requirements regarding drinking water quality and wastewater treatment in rural areas. Approximately 27% of the U.S. population lives in areas defined by the Census Bureau as rural. Many rural communities need to complete water and waste disposal projects to improve the public health and environmental conditions of their citizens. Funding needs are high (more than $88 billion, according to state surveys).

Several federal programs assist rural communities in meeting these requirements. In dollar terms, the largest are administered by the Environmental Protection Agency, but they do not focus solely on rural areas. The Department of Agriculture's grant and loan programs support significant financial activity and are directed solely at rural areas. Meeting infrastructure funding needs of rural areas efficiently and effectively is likely to remain an issue of considerable congressional interest.

* This is an edited, reformatted and augmented version of a Congressional Research Service publication 98-64, prepared for Members and Committees of Congress, dated November 5, 2014.

INTRODUCTION

The public health and environmental requirements of two federal laws are primarily driving projects in rural, as well as non-rural, areas for drinking water and wastewater treatment. The Environmental Protection Agency (EPA) administers both laws.

For *the quality of drinking water supply*, requirements of the Safe Drinking Water Act (SDWA) apply to public water supply systems, whether government-owned or privately owned. Under this law, EPA regulates the quality of drinking water provided by community water supply systems, which are defined as those having at least 15 service connections. Community water supply systems serve approximately 300 million persons; 19 million persons also get their drinking water from non-community systems (such as wells that serve individual homes, schools, factories, or campgrounds), which are not subject to the act.

Regulated water systems provide drinking water to 97% of Americans. The vast majority of systems are small and privately owned, although most people (70% of those served by community water systems) are customers of large, publicly owned systems. The smallest water systems (serving fewer than 3,300 persons, many serving small clusters of homes) account for 77% of all systems and a similarly high percentage of systems in significant noncompliance with drinking water regulations. Most very small systems have no credit history and have never raised capital in financial markets. Small to medium systems (serving 3,301 to 50,000 persons) are institutionally more capable than smaller systems, yet they also face financing challenges. The smallest of these have limited access to financial markets and creditworthiness more sensitive to local economic conditions than larger systems.

Community water supply systems currently are subject to a number of drinking water regulations issued by EPA under the SDWA. Federal regulations limiting levels of contaminants in treated water are implemented by local water suppliers. These require, for example, system monitoring, treatment to remove certain contaminants, and reporting. New regulations are being developed that are likely to impose additional compliance burdens on these systems within the next few years, and costs of meeting these requirements are a growing concern to water suppliers and policy makers.

EPA estimates that compliance with the regulations already promulgated will provide millions of people protection from numerous industrial chemicals, microbes, and other contaminants in public water supplies. However, to comply, many cities and towns must invest in capital equipment, operation and maintenance, and increased staff technical capacity. Recent regulations with particularly costly implications for small towns include water filtration, lead control, arsenic control, and inorganic and organic contaminant control.

According to EPA, small community water systems (serving up to 3,300 persons) have funding needs of $64.5 billion (17% of the total national need) to provide safe drinking water— accounting for a disproportionate percentage of community water system need. Although small systems account for only 8% of the population served, nearly 83% of all systems with reported funding needs are small communities, EPA said.[1]

For *wastewater treatment*, requirements of the Clean Water Act (CWA) apply to all communities that discharge municipal sewage waste into the nation's waters. About 69% of wastewater treatment and collection *facilities* serve small communities (defined as those with a population of 10,000 or less), yet those facilities only serve 10% of the U.S. *population*. Under the CWA, all municipalities were to achieve secondary treatment of municipal sewage (equivalent to removing approximately 85% of wastes from the municipal wastestream), or more stringent, where that is required to meet local water quality standards, by 1988. Much like the Safe Drinking Water Act, small community wastewater systems have higher rates of noncompliance than larger systems. Unlike the SDWA, however, regulatory requirements under the CWA have been fixed for some time. The CWA issue for many cities of all size is continuing efforts to finance improvements that have largely been known for several years.

EPA reported in 2010 that cities throughout the country (of all sizes) would require nearly $298 billion for wastewater facilities to meet water quality standards.[2] EPA estimated that funding needs for rural communities totaled $109 billion. Needs for projects in small communities (populations less than 10,000) were $23.7 billion, or about 8% of the total U.S. funding needs. The largest needs in small communities are for pipe repair and new sewer pipes, improved wastewater treatment, and correction of combined sewer overflows. Seven states accounted for 50% of the small community needs (Pennsylvania, New York, Iowa, Utah, Illinois, West Virginia, and Ohio). With few exceptions, small community facilities are a large majority of the total number of publicly owned facilities in each state, and in nearly one-quarter of the states, small communities facilities constitute more than 80% of publicly owned facilities. Facilities in the smallest communities (fewer than 1,000 persons) represent 27% of small community facility needs. As with meeting drinking water needs, EPA has estimated that, because small systems lack economies of scale, their customers face a particularly heavy financial burden to meet needs for clean water investments. The smallest cities are likely to experience the largest overall percentage increases in user charges and fees as a result, EPA has said.

FEDERAL ASSISTANCE PROGRAMS

The federal government administers a number of programs that assist rural communities in developing water and waste disposal systems. The most prominent are programs of the Department of Housing and Urban Development (HUD); the Appalachian Regional Commission (ARC); the Economic Development Administration (EDA); EPA; and the U.S. Department of Agriculture (USDA).[3]

HUD administers assistance primarily under the Community Development Block Grant (CDBG) program, in Title I of the Housing and Community Development Act of 1974, as amended. FY2014 appropriations were $3.03 billion, of which approximately $907 million is available for smaller communities. CDBG funds are used by localities for a broad range of activities intended to result in decent housing in a suitable living environment. Water and waste disposal needs compete with many other public activities for this assistance and are estimated to account for 10% to 20% of CDBG obligations. Program policy requires that at least 70% of funds must benefit low/moderate-income persons. According to data from HUD, in recent years, water and sewer improvement projects have accounted for 9%-10% of all CDBG funds nationally. Thirty percent of CDBG funds are allocated by formula to the states for distribution to small communities and may be available for rural community projects. The larger portion of total CDBG funds, 70%, is allocated by formula to metropolitan areas and cities with populations of 50,000 or more and statutorily defined urban counties and thus does not assist rural areas directly.

The ARC assists with programs and projects to provide basic facilities essential to economic growth in the Appalachian regions of 13 states. Investments are concentrated in areas with significant potential for future growth as well as in areas that suffer the greatest distress. States recommend projects for assistance. In FY2014, the ARC was funded at a level of $80.3 million, budgeted primarily for area development assistance, covering a range of community-based projects including basic infrastructure, business, and human development. Historically, environmental projects have received about 5% of ARC economic and human development funds.

EDA provides project grants for construction of public facilities, including water and sewer systems, to alleviate unemployment and underemployment in economically distressed areas. Development grants provide for infrastructure projects that foster industries and commercial businesses that provide long-term employment and are part of approved overall economic development programs in areas of lagging economic growth. Economic development grants can be used for a wide range of purposes and frequently have a sewer or water supply element. In FY2014, EDA's public works grants were funded at $96 million.

In historic terms, the largest federal program for wastewater treatment assistance is administered by EPA under the Clean Water Act. Since 1973 Congress has appropriated $91 billion in assistance under this act. FY2014 funding is $1.45 billion. Funds are distributed to states under a statutory allocation formula and are used to assist qualified projects on a priority list that is determined by individual states. Prior to 1989, states used their allotments to make grants to cities and other localities. Now, however, federal funds are used to capitalize state loan programs (State Revolving Funds, or SRFs), and project loans are made according to criteria in the CWA. Over the long term, the loan programs are intended to be sustained through repayment of loans to states, thus creating a continuing source of assistance for other communities. Rural and non-rural communities compete for funding; rural areas have no special priority, nor are states required to reserve any specific percentage for projects in rural areas.

Some small communities and states with large rural populations have had problems with the CWA loan program. Many small towns did not participate in the previous grants program and are more likely to require major projects to achieve compliance with the law. Yet many have limited financial, technical, and legal resources and have encountered difficulties in qualifying for and repaying loans. They often lack an industrial tax base or opportunities for economies of scale and thus face the prospect of very high per capita user fees to repay a loan for the full cost of sewage treatment projects. Still, small communities have been participating in the clean water SRF program: an estimated 23% of the $42 billion in SRF assistance since 1989 (representing 67% of loans made by states) has gone to communities with less than 10,000 in population.

In 1996, Congress enacted Safe Drinking Water Act (SDWA) amendments which authorize federal capitalization of state loan programs to help public water systems finance improvements needed to comply with federal drinking water regulations (P.L. 104-182). This drinking water SRF program (DWSRF) is similar in structure to the CWA SRF program. Elements that differ under the DWSRF include authority for states to make both loans and grants, to assist both privately and publicly owned community water systems, and to provide loan subsidies to economically disadvantaged communities. To give states flexibility in meeting infrastructure needs, the law allows a governor to transfer as much as 33% of the annual DWSRF allotment to the CWA SRF, or an equivalent amount from the CWA SRF to the DWSRF. For FY2014, Congress appropriated $907 million for SDWA SRF assistance.[4] According to EPA, 38% of total assistance since 1996 (representing 71% of all assistance agreements) has gone to systems serving fewer than 10,000 persons.

USDA WATER AND WASTE DISPOSAL PROGRAMS

Grants and loans (direct and guaranteed) for water and wastewater projects are available through rural development programs of the U.S. Department of Agriculture (USDA). Funds are limited to communities with population of 10,000 or less. Communities must be denied credit through normal commercial channels to be eligible for assistance. USDA prefers making loans; grants are made only when necessary to reduce average annual user charges to a reasonable level. The split between loans and grants is about 70-30. In recent years, approximately 65% of loan funds and 57% of grant funds have been obligated to drinking water projects; the remainder have been obligated to waste disposal projects. USDA also makes grants to nonprofit organizations to provide technical assistance and training to assist rural communities with their drinking water, wastewater, and solid waste disposal problems.

Prior to the 1996 farm bill (the Federal Agriculture Improvement and Reform Act of 1996, P.L. 104-127), these USDA grants and loans, as well as other USDA rural development assistance to businesses, industries, and communities, were authorized as separate programs. In P.L. 104-127, Congress endorsed an Administration proposal to consolidate 14 existing rural development grant and loan programs into three categories for better coordination and greater local involvement. The program is called the Rural Community Advancement Program (RCAP). The three components are the Rural Utilities Service (RUS, providing assistance for water and wastewater disposal, solid waste management, and emergency community water programs), Rural Community Facilities, and Rural Business and Cooperative Development.

Under RCAP, USDA state offices work with state and local governments, Indian tribes, and private and community organizations to prepare a strategic plan for delivering RCAP assistance to each state. The key concept in RCAP is to involve state and local stakeholders in strategic planning, so that federal assistance will address local priorities more effectively.

The 1996 farm bill did not alter the basic features or statutory requirements of the water and waste disposal grant and loan programs, which are administered through a network of state and local offices. USDA headquarters allocates program funds to the Rural Economic and Community Development state offices through an allocation formula based on rural population, poverty, and unemployment. District RECD offices actually administer the programs. Since 2001 USDA has provided more than $10 billion to more than 7,500 rural water and wastewater systems, benefitting more than 6.5 million people. Subsequent farm bills in 2002 (P.L. 107-171), 2008 (P.L. 110-246), and 2014 (P.L. 113-79) have not significantly modified USDA's rural water and waste disposal assistance programs. The 2014 law extended authorizations for grants and loans through FY2018.[5]

Water and Waste Disposal Grants

Grants for the development costs of water supply and waste disposal projects in rural areas are authorized under the Consolidated Farm and Rural Development Act. An eligible project must serve a rural area which is not likely to decline in population below that for which the project was designed, and it must be designed and constructed so that adequate capacity will or can be made available to serve the reasonably foreseeable growth needs of the area.

Grants may not exceed 75% of the development cost of a project and should only be used to reduce user costs to a reasonable level. Grants are only made after a determination of the maximum amount of loan that a community can afford and still have reasonable user rates. Grants, which typically provide 35%-45% of project costs, may be used to supplement other funds borrowed or furnished by applicants for project costs and may be combined with loans when the applicant is able to repay part, but not all, of the project costs. Eligible applicants may include municipalities, authorities, districts, certain

Indian tribes, and similar organizations. Priority is given to projects serving populations of fewer than 5,500 persons.

Emergency and Imminent Community Water Assistance Grants

RUS also is authorized to help rural residents where a significant decline in quantity or quality of drinking water exists or is imminent and funds are needed to obtain adequate quantities of water that meet standards of the Safe Drinking Water Act or the Clean Water Act. Grants, ranging from $10,000 to a maximum of $500,000, are provided for projects to serve a rural area with a population of 10,000 or less that has a median household income not in excess of the statewide nonmetropolitan median household income. Grants for repairs, partial replacement, or significant maintenance of an established system cannot exceed $150,000. Communities use the funds for new systems, waterline extensions, construction of water source and treatment facilities, and repairs or renovation of existing systems and may be awarded for 100% of project cost. Applicants compete on a national basis for available funding. The 2014 farm bill authorized $30 million per year through FY2018 for this program, subject to appropriations. Funding for it is provided through reservation of 3% to 5% of appropriated water and waste disposal grant funds. Amounts actually provided through this program have been quite variable over time, depending on need. In FY2012, $3.1 million was distributed in eight states; in FY2013, $5.8 million was distributed in 11 states.

Water and Waste Disposal Loans

The Rural Development Act of 1972 authorized the Rural Development Insurance Fund under the Consolidated Farm and Rural Development Act. Among other activities, this fund is used for loans to develop storage, treatment, purification, or distribution of water or collection, treatment, or disposal of waste in low-income rural areas.

Loans are made to public bodies, not-for-profit organizations, Indian tribes on federal and state reservations, and other federally recognized tribes for projects needed to meet health or sanitary standards, including clean water standards and SDWA requirements. Loans are repayable in not more than 40 years or the useful life of the facility, whichever is less. USDA makes either direct loans to applicants or guarantees up to 90% of loans made by third-party lenders. Borrowers are required to refinance (graduate) to other credit when they can obtain the needed funds from commercial sources at reasonable rates and terms.

Loan interest rates are based on the community's economic and health environment and are designated poverty, market, or intermediate. Poverty interest rate loans are made in areas where the median household income (MHI) falls below the higher of 80% of the statewide nonurban MHI, or the poverty level, and the project is needed to meet health or sanitary standards; by law, this rate is set at 60% of the market rate. The market rate is adjusted quarterly and is set using the average of a specified 11-bond index. It applies to loans to applicants where the MHI of the service area exceeds the statewide nonurban MHI. The intermediate rate applies to loans that do not meet the criteria for the poverty rate and which do not have to pay the market rate; by law, this rate is set at 80% of the market rate.[6] Interest rates on guaranteed loans are negotiated between the borrower and the lender. The 2014 farm bill amended the water and waste disposal direct and guaranteed loan programs to encourage financing by private or cooperative lenders to the maximum extent possible, use of loan guarantees where the population exceeds 5,500, and use of direct loans where the impact of a guaranteed loan on rate payers would be significant.

Beginning with USDA's FY1996 appropriation, Congress consolidated the water and waste disposal grant and loan appropriations in a single Rural Utilities Assistance Program, consistent with the approach taken in the 1996 farm bill to consolidate delivery of rural development assistance. Funds available through appropriations for USDA's water and waste disposal programs (P.L. 113-76) and mandatory funds through the 2014 farm bill (P.L. 113-79) provide $602.4 million in total for FY2014, including $396 million in grants and subsidy to support direct and guaranteed loans. USDA estimated that, counting both appropriations and repaid loan monies still available, those funds would support more than $1 billion in program activity.

In dollar terms, the largest federal programs that solely assist water and waste disposal needs are administered by EPA. Funds available in FY2014 for EPA's clean water and safe drinking water SRF programs total $2.3 billion. They do not focus solely on rural areas, however. USDA's grant and loan programs also support significant financial activity and are directed entirely at rural areas. Still, funding needs in rural areas are high (more than $88 billion, according to state surveys summarized in EPA reports), and there is heavy demand for funds. At the end of FY2007, USDA reported a $2.4 billion backlog of requests for 928 water and wastewater projects for its grant and loan programs.

Among policy makers, interest has grown in identifying or developing alternative financing tools beyond federal grants or SRFs to help communities meet their water infrastructure needs. Finding consensus on an approach and the revenues to support such needed investments are significant challenges. In 2014, Congress enacted legislation that authorizes a five-year pilot program to provide federal loan assistance for water infrastructure projects, called the Water Infrastructure Finance and Innovation Act (WIFIA) program.[7] However, some analysts are concerned that alternative financing approaches—such as the WIFIA pilot program—are more focused on large communities than rural areas, which could have more difficulty making needed infrastructure investments in the future. Meeting the infrastructure funding needs of rural areas efficiently and effectively is likely to remain an issue of considerable congressional interest.

End Notes

[1] U.S. Environmental Protection Agency, Drinking Water Infrastructure Needs Survey and Assessment, Fifth Report to Congress (EPA 816-R-13-006), Washington, DC, April 2013, p. 9.

[2] U.S. Environmental Protection Agency, Clean Watersheds Needs Survey 2008, Report to Congress, Washington, June 2010, http://www.epa.gov/cwns/cwns2008rtc.pdf.

[3] For information, see CRS Report RL30478, Federally Supported Water Supply and Wastewater Treatment Programs, coordinated by Claudia Copeland.

[4] For information, see CRS Report RS22037, Drinking Water State Revolving Fund (DWSRF): Program Overview and Issues, by Mary Tiemann.

[5] For additional information, see CRS Report R43718, Rural Development Provisions in the 2014 Farm Bill (P.L. 113- 79), by Tadlock Cowan.

[6] For current interest rates, see http://www.rurdev.usda.gov/UWP-int-rate.htm.

[7] For information, see CRS Report R43315, Water Infrastructure Financing: The Water Infrastructure Finance and Innovation Act (WIFIA) Program, by Claudia Copeland.

In: Rural Water Systems
Editor: Gloria Simmons

ISBN: 978-1-63483-240-3

Chapter 8

SAFE DRINKING WATER ACT (SDWA): A SUMMARY OF THE ACT AND ITS MAJOR REQUIREMENTS*

Mary Tiemann

SUMMARY

This report summarizes the Safe Drinking Water Act (SDWA) and its major programs and regulatory requirements. This report includes the drinking water security provisions added to the SDWA by the Public Health Security and Bioterrorism Preparedness and Response Act of 2002 (P.L. 107-188), and lead reduction provisions as amended by P.L. 111-380. It also outlines amendments made in December 2013 by P.L. 113-64 (H.R. 3588) to explicitly exempt fire hydrants from coverage under the act's lead plumbing restrictions.

The SDWA, Title XIV of the Public Health Service Act, is the key federal law for protecting public water supplies from harmful contaminants. First enacted in 1974 and substantially amended in 1986 and 1996, the act is administered through programs that establish standards and treatment requirements for public water supplies, control underground injection of wastes, finance infrastructure projects, and protect sources of drinking water. The 1974 law established the current

* This is an edited, reformatted and augmented version of a Congressional Research Service publication, CRS Report for Congress RL31243, prepared for Members and Committees of Congress, from www.crs.gov, dated February 5, 2014.

federal-state arrangement in which states may be delegated primary implementation and enforcement authority for the drinking water program. The state-administered Public Water Supply Supervision (PWSS) Program remains the basic program for regulating the nation's public water systems, and 49 states have assumed this authority.

The last major reauthorization of the act was done through the Safe Drinking Water Act Amendments of 1996 (P.L. 104-182), which generally authorized appropriations for SDWA programs through FY2003. As with other EPA-administered statutes having expired funding authority, Congress has continued to appropriate funds for the ongoing SDWA programs.

In addition to reviewing key programs and requirements of the SDWA, this report includes statistics on the number and types of regulated public water systems. It also provides a table that lists all major amendments, with the year of enactment and public law number, and that cross-reference sections of the act with the major *U.S. Code* sections of the codified statute.

INTRODUCTION

The Safe Drinking Water Act (SDWA), Title XIV of the Public Health Service Act, is the key federal law for protecting public water supplies from harmful contaminants. First enacted in 1974 and substantially amended in 1986 and 1996, the act is administered through programs that establish standards and treatment requirements for public water supplies, control underground injection of wastes, finance infrastructure projects, and protect sources of drinking water. The 1974 law established the current federal-state arrangement in which states may be delegated primary implementation and enforcement authority for the drinking water program. The state-administered Public Water Supply Supervision (PWSS) Program remains the basic program for regulating the nation's public water systems, and 49 states have assumed this authority. The 1996 SDWA amendments generally authorized appropriations for SDWA programs through FY2003. *Table 1* identifies the original enactment and subsequent amendments.

This report summarizes the act's major provisions, programs, and requirements. It also provides selected statistics on the universe of regulated public water systems, and lists references for further information on the act and its implementation. *Table 2*, located at the end of this report, cites the major *U.S. Code* sections of the act and the equivalent sections of the statute.

Table 1. Safe Drinking Water Act and Amendments (codified generally as 42 U.S.C. 300f-300j)

Year	Act	Public Law Number
1974	Safe Drinking Water Act of 1974	P.L. 93-523
1977	Safe Drinking Water Act Amendments of 1977	P.L. 95-190
1979	Safe Drinking Water Act Amendments	P.L. 96-63
1980	Safe Drinking Water Act Amendments	P.L. 96-502
1986	Safe Drinking Water Act Amendments of 1986	P.L. 99-339
1988	Lead Contamination Control Act of 1988	P.L. 100-572
1996	Safe Drinking Water Act Amendments of 1996	P.L. 104-182
2002	Public Health Security and Bioterrorism Preparedness and Response Act of 2002	P.L. 107-188
2011	Reduction of Lead in Drinking Water Act	P.L. 111-380
2013	Community Fire Safety Act of 2013	P.L. 113-64

BACKGROUND

As indicated by *Table 1*, the Safe Drinking Water Act has been amended several times since enactment of the Safe Drinking Water Act of 1974 (P.L. 93-523). Congress passed this law after nationwide studies of community water systems revealed widespread water quality problems and health risks resulting from poor operating procedures, inadequate facilities, and uneven management of public water supplies in communities of all sizes. The 1974 law gave the EPA substantial discretionary authority to regulate drinking water contaminants and gave states the lead role in implementation and enforcement.

The first major amendments (P.L. 99-339), enacted in 1986, were largely intended to increase the pace at which the EPA regulated contaminants and to increase the protection of ground water. From 1974 until 1986, the EPA had regulated just one additional contaminant beyond the 22 standards previously developed by the Public Health Service. The 1986 amendments required the EPA to (1) issue regulations for 83 specified contaminants by June 1989 and for 25 more contaminants every three years thereafter, (2) promulgate requirements for disinfection and filtration of public water supplies, (3) ban the use of lead pipes and lead solder in new drinking water systems, (4) establish an elective wellhead protection program around public wells, (5) establish a demonstration grant program for state and local authorities having designated sole-source aquifers to develop ground water protection programs, and (6) issue rules for monitoring injection wells that inject wastes below a drinking water source. The amendments also increased EPA's enforcement authority.

Congress again amended SDWA with the Lead Contamination Control Act of 1988 (P.L. 100- 572). These provisions were intended to reduce exposure to lead in drinking water by requiring the recall of lead-lined water coolers, and requiring the EPA to issue a guidance document and testing protocol for states to help schools and day care centers identify and correct lead contamination in school drinking water.

After the regulatory schedule mandated in the 1986 amendments proved to be unworkable for the EPA, states, and public water systems, the 104th Congress made sweeping changes to the act with the SDWA Amendments of 1996 (P.L. 104-182). As over-arching themes, the amendments targeted resources to address the greatest health risks, added some regulatory flexibility, provided funding for federal drinking water mandates, and aimed to improve water system compliance capacity. Congress revoked the requirement that the EPA regulate 25 new contaminants every three years and created a risk-based approach for selecting contaminants for regulation.

Among other changes, Congress added some flexibility to the standard setting process, required the EPA to conduct health risk reduction and cost analyses for new rules, authorized a drinking water state revolving loan fund (DWSRF) program to help water systems finance projects needed to meet SDWA requirements, added programs to improve small system compliance, expanded consumer information requirements, increased the act's focus on pollution prevention through a voluntary source water protection program, and streamlined the act's enforcement provisions. P.L. 104-182 extended authorizations for appropriations under the act through FY2003.

In 2002, several drinking water security provisions were added to the SDWA through the Public Health Security and Bioterrorism Preparedness and Response Act of 2002 (P.L. 107-188). Title IV of the act included requirements for community water systems serving more than 3,300 individuals to conduct vulnerability assessments and prepare emergency response plans. The law also required the EPA to conduct research on preventing and responding to terrorist or other attacks.

REGULATED PUBLIC WATER SYSTEMS

Federal drinking water regulations apply to the approximately 152,700 privately and publicly owned water systems that provide piped water for human consumption to at least 15 service connections or that regularly serve at least 25 people. These water systems vary greatly in size and type, ranging from large municipal systems to homeowner associations, schools, and campgrounds.

More than 51,350 of the regulated public water systems are *community water systems* (CWS) that serve the same residences year-round. These water systems provide water to approximately 299 million people. All federal regulations apply to these systems. Most community water systems (82%) are relatively small, serving 3,300 people or fewer. Despite this large percentage, these systems provide water to just 9% of the total population served by community water systems. In contrast, 8% of CWS serve populations of 10,000 or more, but provide water to 82% of the population served (more than 246 million individuals). Among the community water systems, 77% rely on ground water and 23% rely on surface water.

Another 18,178 public water systems are *non-transient non-community water systems* (NTNCWS), such as schools or factories, which have their own water supply and generally serve the same individuals for more than six months but not year-round. Most drinking water regulations apply to these systems. Of these water systems, 99% are small or very small, and provide water to 83% of the population served by these systems.

Nearly 83,200 other public water systems are *transient non-community water systems* (TNCWS), such as campgrounds and gas stations, which provide their own water to transitory customers. Only regulations for contaminants that pose immediate health risks apply to these systems.[1]

Approximately 95,700 of the nearly 101,400 non-community water systems (transient and non-transient systems combined) serve 500 or fewer people. These statistics give some insight into the scope of technological, economic, and managerial challenges small public water systems may face in meeting federal drinking water regulations. (*Figure 1* provides statistics on community water systems, non-transient non-community water systems, and transient non-community water systems.)

NATIONAL DRINKING WATER REGULATIONS

A key component of SDWA is the requirement that the EPA promulgate national primary drinking water regulations for contaminants that may pose health risks and that are likely to be present in public water supplies. Section 1412 instructs the EPA on how to select contaminants for regulation and specifies how the EPA must establish regulations once a contaminant has been selected. The regulations apply to the roughly 168,000 privately and publicly owned water systems that provide piped water for human consumption to at least 15 service connections or that regularly serve at least 25 people. The EPA has issued regulations for more than 90 contaminants, including regulations setting new standards for drinking water disinfectants and their byproducts and for microbial contaminants, a regulation establishing a standard for uranium in drinking water, and a regulation revising the standard for arsenic.

		Water System Population Size Category					
		Very Small 500 or less	Small 501-3,300	Medium 3,301-10,000	Large 10,001-100,000	Very Large >100,000	Totals
CWS	# Systems	28,3462	13,737	4,936	3,802	419	51,356
	Pop. Served	4,763,672	19,661,787	28,737,564	108,770,014	137,283,104	299,216,141
	% of Systems	55%	27%	10%	7%	1%	100%
	% Of Pop.	2%	7%	10%	36%	46%	100%
NTNCWS	# Systems	15,461	2,566	132	18	1	18,178
	Pop. Served	2,164,594	2,674,694	705,320	441,827	203,000	6,189,435
	% of Systems	85%	14%	1%	0%	0%	100%
	% Of Pop.	35%	43%	11%	7%	3%	100%
TNCWS	# Systems	80,347	2,726	92	13	1	83,179
	Pop. Served	7,171,054	2,630,931	514,925	334,715	2,000,000	12,651,625
	% of Systems	97%	3%	0%	0%	0%	100%
	% Of Pop.	57%	21%	4%	3%	16%	100%
	Total # of Systems	124,270	19,029	5,160	3,833	421	152,713

Source: U.S. Environmental Protection Agency, Fiscal Year 2011 Drinking Water and Ground Water Statistics, EPA 816-R-13-003, March 2013, p. 8, http://water.epa.gov/scitech/datait/databases/drink/sdwisfed/upload/epa816r13003.pdf.

Notes: EPA has established three broad categories of public water systems. A community water system (CWS) serves the same population year-round. A non-transient non-community water system (NTNCWS) regularly supplies water to at least 25 of the same people at least six months per year, but not year-round (e.g., schools, factories, office buildings, and hospitals that have their own water systems.

Transient non-community water systems (TNCWS) provide water in places where people do not remain for long periods of time, such as gas stations and campgrounds.

Figure 1. Public Water System Statistics. (water systems regulated under the SDWA)

Contaminant Selection and Regulatory Schedules

Section 1412, as amended in 1996, directs the EPA to select contaminants for regulatory consideration based on occurrence, health effects, and meaningful opportunity for health risk reduction. Starting in 1998 and then every five years, the EPA must publish a list of contaminants that may warrant regulation. As of 2001, and every five years thereafter, the EPA must determine whether or not to regulate at least five of the listed contaminants. The act requires the EPA to evaluate contaminants that present the greatest health concern and to regulate those contaminants that occur at concentration levels and frequencies of public health concern. The amendments also included schedules for the EPA to complete regulations for specific contaminants (i.e., radon, arsenic, disinfectants and disinfection byproducts and *Cryptosporidium*).

Standard Setting

For each contaminant that the EPA determines requires regulation, the EPA must set a nonenforceable maximum contaminant level goal (MCLG) at a level at which no known or anticipated adverse health effects occur and which allows an adequate margin of safety. The EPA must then set an enforceable standard, a maximum contaminant level (MCL), as close to the MCLG as is "feasible" using best technology, treatment techniques, or other means available (taking costs into consideration). The agency generally sets standards based on technologies that are affordable for large communities; however, as amended by P.L. 104-182, the act requires EPA, when issuing a regulation for a contaminant, to list any technologies or other means that comply with the MCL and that are affordable for small public water systems serving populations of 10,000 or fewer. If the EPA does not identify "compliance" technologies that are affordable for small systems, then it must identify small system "variance" technologies or other means that may not achieve the MCL but are protective of public health. The 1996 amendments authorized the EPA to set a standard at other than the feasible level if the feasible level would lead to an increase in health risks by increasing the concentration of other contaminants or by interfering with the treatment processes used to comply with other SDWA regulations. In such cases, the standard or treatment techniques must minimize the overall health risk. Also, when proposing a regulation, the EPA now must publish a determination as to whether or not the benefits of the standard justify the costs. If the EPA determines that the benefits do not justify the costs, the agency may, with certain exceptions, promulgate a standard that maximizes health risk reduction benefits at a cost that is justified by the benefits.[2]

New regulations generally become effective three years after promulgation. Up to two additional years may be allowed if the EPA (or a state in the case of an individual system) determines the time is needed for capital improvements. (Section 1448 outlines procedures for judicial review of EPA actions involving the establishment of SDWA regulations and other final EPA actions.)

Risk Assessment

The 1996 amendments also added risk assessment and risk communication provisions to SDWA. When developing regulations, the EPA is required to (1) use the best available, peer-reviewed science and supporting studies and data; and (2) make publicly available a risk assessment document that discusses estimated risks, uncertainties, and studies used in the assessment. When proposing drinking water regulations, the EPA must publish a health risk reduction and cost analysis (HRRCA). The EPA may promulgate an interim standard without first preparing this benefit-cost analysis or making a determination as to whether the benefits of a regulation would justify the costs if the Administrator determines that a contaminant presents an urgent threat to public health.

Variances and Exemptions

In anticipation that some systems, particularly smaller ones, could have difficulty complying with every regulation, Congress included in SDWA provisions for variances and exemptions. Section 1415 authorizes a state to grant a public water system a *variance* from a standard if raw water quality prevents meeting the standard despite application of best technology, and the variance does not result in an unreasonable risk to health.

Subsection 1415(e) authorizes variances specifically for small systems, based on application of best affordable technology. When developing a regulation, if the EPA cannot identify a technology that meets the standard and is affordable for small systems, the EPA must identify variance technologies that are affordable but do not necessarily meet the standard. In cases where the EPA has identified variance technologies, then states may grant small system variances to systems serving 3,300 or fewer persons if the system cannot afford to comply with a standard (through treatment, an alternative water source, or restructuring) and the variance ensures adequate protection of public health. A state may then grant a variance to a small system, allowing the system to use a variance technology to comply with a regulation. With EPA approval, states also may grant variances to systems serving between 3,301 and 10,000 persons. Variances are not available for microbial contaminants. *The EPA has determined that affordable compliance technologies are available for all existing standards. Thus, small system variances are not available.*

Section 1416 authorizes states to grant public water systems temporary *exemptions* from standards or treatment techniques if a system cannot comply for other compelling reasons (including costs). An exemption is intended to give a water system more time to comply with a regulation and can be issued only if it will not result in an unreasonable health risk. A qualified system may receive an exemption for up to three years beyond the compliance deadline. Systems serving 3,300 or fewer persons may receive a maximum of three additional two-year extensions, for a total exemption duration of nine years.

STATE PRIMACY

Section 1413 authorizes states to assume primary oversight and enforcement responsibility (primacy) for public water systems. To assume primacy, states must adopt regulations at least as stringent as national requirements, develop adequate procedures for enforcement (including conducting monitoring and inspections), adopt authority for administrative penalties, and maintain records and make reports as EPA may require. States also must develop a plan for providing safe drinking water under emergency circumstances. Currently, 55 of 57 states and territories have primacy authority for the public water system supervision (PWSS) program.

Under Section 1443, Congress authorized appropriations of $100 million annually for the EPA to make grants to states to administer the PWSS program. This section directs the EPA, in accordance with regulations, to allot the sums among the states "on the basis of population, geographical area, number of public water systems, and other relevant factors." States are authorized to use a portion of their drinking water state revolving fund grant (under Section 1452) to cover the costs of administering the PWSS program.

ENFORCEMENT, CONSUMER INFORMATION, AND CITIZEN SUITS

The Safe Drinking Water Act requires public water systems to monitor their water supplies to ensure compliance with drinking water standards and to report monitoring results to the states. States review monitoring data submitted by public water systems, and also conduct their own monitoring, to determine system compliance with drinking water regulations. The EPA monitors public water system compliance primarily by reviewing the violation data submitted by the states.

Section 1414 requires that, whenever the EPA finds that a public water system in a state with primary enforcement authority does not comply with regulations, the agency must notify the state and the system and provide assistance to bring the system into compliance. If the state fails to commence enforcement action within 30 days after the notification, the EPA is authorized to issue an administrative order or commence a civil action. In a nonprimacy state, the EPA must notify an elected local official (if any has jurisdiction over the water system) before commencing an enforcement action against the system.

The 1996 amendments strengthened enforcement authorities, streamlined the process for issuing federal administrative orders, increased administrative penalty amounts, made more sections of the act clearly subject to EPA enforcement, and required states (as a condition of primacy) to have administrative penalty authority. The amendments also provided that no enforcement action may be taken against a public water system that has a plan to consolidate with another system.[3]

Consumer Information and Reports

Enforcement provisions also require public water systems to notify customers of violations of drinking water standards or other requirements, such as monitoring and reporting requirements. Systems must notify customers within 24 hours of any violations that have the potential to cause serious health effects. Additionally, community water systems must mail to all customers an annual "consumer confidence report" on contaminants detected in their drinking water. States are required to prepare annual reports on the compliance of public water systems and to make summaries available to the EPA and the public; the EPA must prepare annual national compliance reports (§1414(c)).

Citizen Suits

Section 1449 provides for citizens' civil actions. Citizen suits may be brought against any person or agency allegedly in violation of provisions of the act, or against the EPA Administrator for alleged failure to perform any action or duty which is not discretionary.

COMPLIANCE IMPROVEMENT PROGRAMS

The 1996 amendments added two state-administered programs aimed at improving public water system compliance with drinking water regulations: the operator certification program and the capacity development program. Section 1419 required states to adopt programs for training and certifying operators of community and non-transient non-community systems (e.g., schools and workplaces that have their own wells). The EPA is required to withhold 20% of a state's annual DWSRF grant, unless the state has adopted and implements an operator certification program. Relatedly, Section 1420 required states to establish capacity development programs, also based on EPA guidance. Congress specified that the programs must include (1) legal authority to ensure that new systems have the technical, financial, and managerial capacity to meet SDWA requirements; and (2) a strategy to assist existing systems that are experiencing difficulties to come into compliance. The EPA is required to withhold a portion of SRF grants from states that do not have capacity development strategies. The agency has not had to withhold funds under either of these programs.

GROUND WATER PROTECTION PROGRAMS

Most public water systems rely on ground water as a source of drinking water, and Part C of the act focuses on ground water protection.[4] Section 1421 authorized the establishment of state underground injection control (UIC) programs to protect underground sources of drinking water (USDWs). In 1977, the EPA issued mandated regulations that contained minimum requirements for state UIC programs to prevent underground injection that endangers drinking water sources, and that required states to prohibit any underground injection not authorized by state permit. The law specified that the regulations could not interfere with the underground injection of brine from oil and gas production or recovery of oil unless underground sources of drinking water would be affected.[5]

Section 1422 authorized affected states to submit plans to the EPA for implementing UIC programs and, if approved, to assume primary enforcement responsibility. If a state's plan has not been approved, or the state has chosen not to assume program responsibility, then the EPA must implement the program (§1423). For oil and gas injection operations only, states with UIC programs are delegated primary enforcement authority without meeting EPA regulations under Section 1421, provided states demonstrate that they have an effective program that prevents underground injection that endangers drinking water sources (§1425, added in 1980). EPA has delegated primacy for all classes of wells to 34 states; it shares implementation responsibility in six states, and implements the UIC program for all well classes in 10 states.

To implement this program, EPA placed UIC wells in five classes based on similarity in the fluids injected, and construction, injection depth, design, and operating techniques, and issued regulations that establish performance criteria for each class.[6] In 2008, the agency proposed a rule to establish a new Class VI well for use in the underground injection of carbon dioxide. Class VI wells would be used for the long-term geologic sequestration of carbon dioxide as a tool for mitigating greenhouse gas emissions from coal-fired power plants and other large stationary sources of carbon dioxide.

A separate ground water protection provision, Subsection 1424(e), authorizes the EPA to make determinations, on EPA's initiative or upon petition, that an aquifer is the sole or principal drinking water source for an area. In areas that overlie a designated sole-source aquifer, no federal funding may be committed for projects that the EPA determines may contaminate such an aquifer. Any person may petition for sole source aquifer designation. Nationwide, the EPA had designated 73 sole source aquifers.

The act contains three other state programs aimed specifically at protecting ground water. Added in 1986, Section 1427 established procedures for demonstration programs to develop, implement, and assess critical aquifer protection areas already designated by the Administrator as sole source aquifers. Section 1428, also added in 1986, established an elective state program for protecting wellhead areas around public water system wells. If a state established a wellhead protection program by 1989, and the EPA approved the state's program, then the EPA may award grants covering between 50% and 90% of the costs of implementing the program. In 1996, Congress added Section 1429, authorizing the EPA to make 50% grants to states to develop programs to ensure coordinated and comprehensive protection of ground water within the states. For these programs, appropriations were authorized through FY2003 as follows: $15 million per year for Section 1427, $30 million per year for Section 1428, and $15 million per year for Section 1429.

SOURCE WATER ASSESSMENT AND PROTECTION PROGRAMS

The 1996 amendments expanded the act's pollution prevention focus to embrace protection of surface water, as well as ground water. Section 1453 required the EPA to publish guidance for states to implement source water assessment programs that delineate boundaries of the areas from which systems receive water, and identify the origins of contaminants in those areas to determine systems' susceptibility to contamination. States with approved assessment programs may adopt alternative monitoring requirements for water systems, as provided for in Section 1418.

Section 1454 authorized a source water petition program based on voluntary partnerships between state and local governments. States may establish a program under which a community water system or local government may submit a petition to the state requesting assistance in developing a voluntary source water quality protection partnership to (1) reduce the presence of contaminants in drinking water, (2) receive financial or technical assistance, and (3) develop a long-term source water protection strategy. This section authorized $5 million each year for grants to states to support petition programs. Also, states may use up to 10% of their DWSRF grant to support various source water protection activities including the petition program.

State Revolving Funds

In 1996, Congress authorized a drinking water state revolving loan fund (DWSRF) program to help systems finance improvements needed to comply with SDWA regulations (§1452). The EPA is authorized to make grants to states to capitalize DWSRFs, which states then may use to make loans to public water systems. States must match 20% of the federal grant. Grants are allotted based on the results of needs surveys.[7] Each state and the District of Columbia must receive at least 1% of the appropriated funds.

Drinking water SRFs may be used to provide loans for expenditures that the EPA has determined will facilitate compliance or significantly further the act's health protection objectives. States must make available 15% of their annual allotment for loan assistance to systems that serve 10,000 or fewer persons, to the extent that funds can be obligated for eligible projects. States may use up to 30% of their DWSRF grant to provide loan subsidies (including forgiveness of principal) to help economically disadvantaged communities. Also, states may use a portion of funds for technical assistance, source water protection and capacity development programs, and for operator certification.

The law authorized appropriations of $599 million for FY1994 and $1 billion per year for FY1995 through FY2003 for DWSRF capitalization grants. The EPA was directed to reserve, from annual DWSRF appropriations, 0.33% for financial assistance to several Trusts and Territories, $10 million for health effects research on drinking water contaminants, $2 million for the costs of monitoring for unregulated contaminants, and up to 2% for technical assistance. The EPA may use 1.5% of funds each year for making grants to Indian Tribes and Alaska Native villages.[8]

Drinking Water Security

The 107th Congress passed the Public Health Security and Bioterrorism Preparedness and Response Act of 2002 (P.L. 107-188) to address a wide range of security issues. Title IV of the Bioterrorism Act amended SDWA to address threats to drinking water security. Key provisions are summarized below.

Vulnerability Assessments

Under new SDWA Section 1433, each community water system serving more than 3,300 people was required to conduct an assessment of the system's vulnerability to terrorist attacks or other intentional acts intended to disrupt the provision of a safe and reliable drinking water supply. This provision established deadlines, based on system size, for community water systems to certify to the EPA that they had conducted a vulnerability assessment and to submit to the EPA a copy of the assessment. Section 1433 exempted the contents of the vulnerability assessments from disclosure under the Freedom of Information Act (except for information contained in the certification identifying the system and the date of the certification), and it provides for civil and criminal penalties for inappropriate disclosure of information by government officials.

Section 1433 further required each of these community water systems to prepare an emergency response plan incorporating the results of the vulnerability assessment. The EPA was directed to provide guidance to smaller systems on how to conduct vulnerability assessments, prepare emergency response plans, and address threats.

Section 1433(e) authorized the appropriation of $160 million for FY2002, and such sums as may be necessary for FY2003 through FY2005, to provide financial assistance to community water systems to conduct vulnerability assessments, to prepare response plans, and for expenses and contracts to address basic security enhancements and significant threats.

The Bioterrorism Act also added Sections 1434 and 1435 to SDWA, directing the Administrator to review methods by which terrorists or others could contaminate or otherwise disrupt the provision of safe water supplies. These provisions require the EPA to review methods for preventing, detecting, and responding to such disruptions, and methods for providing alternative drinking water supplies if a water system was destroyed or impaired. Section 1435(e) authorized $15 million for FY2002, and such sums as may be necessary for FY2003 through FY2005 to carry out Sections 1434 and 1435.

Emergency Powers

Under Section 1431, the Administrator has emergency powers to issue orders and commence civil action if (1) a contaminant likely to enter a public drinking water supply system poses a substantial threat to public health, and (2) state or local officials have not taken adequate action. The Bioterrorism Act amended this section to specify that EPA's emergency powers include the authority to act when there is a threatened or potential terrorist attack or other intentional act to disrupt the provision of safe drinking water or to impact the safety of a community's drinking water supply.

Tampering with Public Water Systems

Section 1432 provides for civil and criminal penalties against any person who tampers, attempts to tamper, or makes a threat to tamper with a public water system. Amendments made by the Bioterrorism Act increased criminal and civil penalties for tampering, attempting to tamper, or making threats to tamper with public water supplies. The maximum prison sentence for tampering increased from 5 to 20 years. The maximum prison sentence for attempting to tamper, or making threats to tamper, increased from 3 to 10 years. The maximum fine that may be imposed for tampering increased from $50,000 to $1 million. The maximum fine for attempting to tamper, or threatening to tamper, increased from $20,000 to $100,000.

Emergency Assistance

SDWA Subsection 1442(b) authorizes the EPA to provide technical assistance and to make grants to states and public water systems to assist in responding to and alleviating emergency situations. The Bioterrorism Act amended Subsection 1442(d) to authorize appropriations for such emergency assistance of not more than $35 million for FY2002, and such sums as may be necessary for each fiscal year thereafter.

ADDITIONAL SDWA PROVISIONS

Lead-Free Plumbing

Section 1417 broadly prohibits the use of any pipe, any pipe or plumbing fitting or fixture, solder, or flux in the installation or repair of public water systems or plumbing in residential or nonresidential facilities providing drinking water that is not "lead free" (as defined in the act). This section also makes it unlawful to sell solder or flux that is not lead-free (unless it is properly labeled), or pipes, plumbing fittings, or fixtures that are not lead-free, with the exception of pipes used in manufacturing or industrial processing or other specific exceptions.[9] Added in 1986, Section 1417(d) defined "lead free" to mean not more than 0.2% lead for solders and fluxes, and not more than 8% lead for pipes and pipe fittings. In 1996, Congress added Section 1417(e), directing EPA to issue regulations setting health-based performance standards limiting the amount of lead that may leach from new plumbing fittings and fixtures unless a voluntary industry standard was established within one year of enactment. An industry standard was established.

Enacted January 4, 2011, the Reduction of Lead in Drinking Water Act (P.L. 111-380) amended Section 1417 to revise the SDWA definition of "lead free" and to add new exemptions from prohibitions on the use or sale of lead pipes, plumbing, and fittings and fixtures. The act reduced the allowable level of lead in products in contact with drinking water from 8.0% to 0.25% (weighted average),[10] and exempted from the general prohibitions "(A) pipes, pipe fittings, plumbing fittings, or fixtures, including backflow preventers, that are *used exclusively for nonpotable services* such as ... industrial processing, irrigation, outdoor watering or any other uses where the water is not anticipated to be used for human consumption"; and (B) various specified products including tub fillers, shower valves, service saddles, or water distribution main gate valves at least 2 inches in diameter. (Italics added.)

P.L. 111-380 removed the reference to Section 1417(e), which required that plumbing fittings and fixtures "intended by the manufacturer to dispense water for human ingestion" must comply with the industry standard; rather, these products became subject to the definition of "lead free" in Section 1417(d). The provisions of P.L. 111-380 became effective on January 4, 2014, and any product that does not meet the 0.25% lead limit may no longer be sold or installed, unless exempt from the general prohibitions.

In October 2013, EPA announced that fire hydrants would not qualify for the exclusion for pipes, fittings, and fixtures used exclusively for non-potable services and, thus, would be required to meet the new lead-free standards.[11] In December 2013, the House and Senate passed H.R. 3588, the Community Fire Safety Act of 2013, to amend the SDWA to clearly exempt fire hydrants from coverage under Section 1417. The President signed the bill into law (P.L. 113-64) on December 20, 2013. P.L. 113-64 also directs EPA to (1) consult with the National Drinking Water Advisory Council on potential revisions to the SDWA regulations for lead, and (2) request the Council to consider lead sources throughout the drinking water distribution system, including components used to reroute water during repairs.[12]

Research, Technical Assistance, and Training

Section 1442 authorizes the EPA to conduct research, studies, and demonstrations related to the causes, treatment, control, and prevention of diseases resulting from contaminants in water. The agency is directed to provide technical assistance to the states and municipalities in administering their public water system regulatory responsibilities. This section authorized $15 million annually for technical assistance to small systems and Indian Tribes, and $25 million for health effects research. (Title II of P.L. 104-182, the 1996 amendments, authorized additional appropriations for drinking water research, not to exceed $26.6 million annually for FY1997 through FY2003.)

Demonstration Grants

The Administrator may make grants to develop and demonstrate new technologies for providing safe drinking water and to investigate health implications involved in the reclamation/reuse of waste waters (§1444).

Records, Inspections, and Monitoring

Section 1445 states that persons subject to requirements under SDWA must establish and maintain records, conduct water monitoring, and provide any information that the Administrator may require by regulation to carry out the requirements of the act. Section 1445(b) authorizes the Administrator or a representative, after notifying the state in writing, to enter and inspect the property of water suppliers or other persons subject to the act's requirements, to determine whether the person is in compliance with the act. Failure to comply with these provisions may result in civil penalties.

This section also directs the EPA to promulgate regulations establishing the criteria for a monitoring program for unregulated contaminants. Beginning in 1999 and every five years thereafter, the EPA must issue a list of not more than 30 unregulated contaminants to be monitored by public water systems. States are permitted to develop representative monitoring plan to assess the occurrence of unregulated contaminants in small systems, and the section authorized $10 million to be appropriated for each of FY1999 through FY2003 to provide grants to cover the costs of monitoring for small systems. All monitoring results are to be included in a national drinking water occurrence data base created under the 1996 amendments.

National Drinking Water Advisory Council

The act established a National Drinking Water Advisory Council, composed of 15 members (with at least two representing rural systems), to advise, consult, and make recommendations to the Administrator on activities and policies derived from the act (§1446).

Federal Agencies

Any federal agency having jurisdiction over federally owned public water systems must comply with all federal, state and local drinking water requirements as well as any underground injection control programs. The act provides for waivers in the interest of national security (§1447).

Assistance to Colonias[13]

Added in 1996, Section 1456 authorized the EPA and other appropriate federal agencies to award grants to Arizona, California, New Mexico, and Texas to provide assistance (not more than 50% of project costs) to *colonias* where the residents are subject to a significant health risk attributable to the lack of access to an adequate and affordable drinking water system. Congress authorized appropriations of $25 million for each of fiscal years 1997 through 1999.

Estrogenic Substances

Section 1547 authorized the EPA to use the estrogenic substances screening program created in the Food Quality Protection Act of 1996 (P.L. 104-170) to provide for testing of substances that may be found in drinking water if the Administrator determines that a substantial population may be exposed to such substances.

Drinking Water Studies

Section 1458 directed the EPA to conduct drinking water studies involving subpopulations at greater risk and biological mechanisms. EPA also was directed to conduct studies to support specific regulations, including those for disinfectants and disinfection byproducts and *Cryptosporidium*.

Selected P.L. 104-182 Provisions Not Amending SDWA

The 104th Congress included a variety of drinking water-related provisions in the 1996 SDWA Amendments that did not amend the Safe Drinking Water Act. Several of these provisions are described below.

Transfer of Funds

Section 302 authorized states to transfer as much as 33% of their annual drinking water state revolving fund grant to the Clean Water Act (CWA) SRF, or an equivalent amount from the CWA SRF to the DWSRF through FY2001.

In several subsequent conference reports for EPA appropriations, Congress authorized states to continue making transfers between the two funds, and in P.L. 109-54, Congress made this authority permanent.[14]

Grants to Alaska

Section 303 of the 1996 amendments authorized the EPA to make grants to the State of Alaska to pay 50% of the costs of improving sanitation for rural and Alaska Native villages. Grants are for construction of public water and wastewater systems, and for training and technical assistance programs. Appropriations were authorized at $15 million for each of fiscal years 1997 through 2000. (In P.L. 106-457, Congress reauthorized appropriations for these rural sanitation grants at a level of $40 million for each of fiscal years 2001 through 2005.)

Bottled Water

Section 305 revised Section 410 of the Federal Food, Drug, and Cosmetic Act to require the Secretary of Health and Human Services to issue bottled drinking water standards for contaminants regulated under SDWA within 180 days after the EPA promulgates the new standards, unless the Secretary determines that a standard is not necessary.

Wastewater Assistance to Colonias

Section 307 authorized EPA to make grants to colonias for wastewater treatment works. Appropriations were authorized at $25 million for each of fiscal years 1997 through 1999.

Additional Infrastructure Funding

Section 401 authorized additional assistance, up to $50 million for each of fiscal years 1997 through 2003, for a grant program for infrastructure and watershed protection projects.

Table 2. U.S. Code Sections of the Safe Drinking Water Act (Title XIV of the Public Health Service Act) (42 U.S.C. 300f-300j-26)

42 U.S.C.	Section Title	SDWA(as amended)
Chapter 6A	Public Health Service	
Subchapter XII	*Safety of Public Drinking Water Systems*	
Part A	Definitions	
300f	Definitions	§1401
Part B	Public Water Systems	
300g	Coverage	§1411
300g-1	National drinking water regulations	§1412
300g-2	State primary enforcement responsibility	§1413
300g-3	Enforcement of drinking water regulations	§1414
300g-4	Variances	§1415
300g-5	Exemptions	§1416
300g-6	Prohibitions on the use of lead pipes, solder, and flux	§1417
300g-7	Monitoring of contaminants	§1418
300g-8a	Operator certification	§1419
300g-9a	Capacity development	§1420
Part C	Protection of Underground Sources of Drinking Water	
300h	Regulations for state programs	§1421
300h-1	State primary enforcement responsibility	§1422
300h-2	Enforcement of program	§1423
300h-3	Interim regulation of underground injections	§1424
300h-4	Optional demonstration by states relating to oil and natural gas	§1425
300h-5	Regulation of state programs	§1426
300h-6a	Sole source aquifer demonstration program	§1427
300h-7a	State programs to establish wellhead protection areas	§1428
300h-8a	State ground water protection grants	§1429
Part D	Emergency Powers	
300i	Emergency powers	§1431
300i-1	Tampering with public water systems	§1432
300i-2a	Terrorist and other intentional acts	§1433
300i-3	Contaminant prevention, detection, and response	§1434
300i-4[a]	Supply disruption prevention, detection and response	§1435

42 U.S.C.	Section Title	SDWA(as amended)
Part E	General Provisions	
300j	Assurance of availability of adequate supplies of chemicals necessary for treatment of water	§1441
300j-1[a]	Research, technical assistance, information, training of personnel	§1442
300j-2[a]	Grants for state programs	§1443
300j-3[a]	Special project grants and guaranteed loans	§1444
300j-4[a]	Records and inspections	§1445
300j-5	National Drinking Water Advisory Council	§1446
300j-6	Federal agencies	§1447
300j-7	Judicial reviews	§1448
300j-8	Citizen civil actions	§1449
300j-9	General provisions	§1450
300j-11	Indian Tribes	§1451
300j-12[a]	State revolving loan funds	§1452
300j-13	Source water quality assessment	§1453
300j-14[a]	Source water petition program	§1454
300j-15	Water conservation plan	§1455
300j-16[a]	Assistance to colonias	§1456
300j-17	Estrogenic substances screening program	§1457
300j-18[a]	Drinking water studies	§1458
Part F	Additional Requirements to Regulate the Safety of Drinking Water	
300j-21	Definitions	§1461
300j-22	Recall of drinking water coolers with lead-lined tanks	§1462
300j-23	Drinking water coolers containing lead	§1463
300j-24	Lead contamination in school drinking water	§1464
300j-25[a]	Federal assistance for state programs regarding lead contamination in school drinking water	§1465
300j-26	Certification of testing laboratories	[b]

Note: This table shows only the major code sections. For more detail and to determine when a section was added, the reader should consult the official printed version of the U.S. Code.

[a] These sections include authorizations of appropriations.

[b] This provision was added by the Lead Contamination Control Act (P.L. 100-572, §4), which did not amend SDWA.

End Notes

[1] EPA's long-standing policy is to exclude transient systems from drinking water regulations except for those contaminants, such as nitrate, that the EPA believes have the potential to cause immediate adverse human health effects resulting from short-term exposure. (Source: Environmental Protection Agency. National Primary Drinking Water Regulation on Lead and Copper, minor revisions. January 12, 2000 (65 Federal Register 1950).)

[2] For a discussion of EPA regulation of contaminants under the revised process, see, for example, U.S. Government Accountability Office, EPA Should Improve Implementation of Requirements on Whether to Regulate Additional Contaminants, GAO-11-254, May 27, 2011, 146 p. http://www.gao.gov/products/GAO-11-254.

[3] For enforcement information and statistics, see U.S. Environmental Protection Agency, Providing Safe Drinking Water in America: 2011 National Public Water Systems Compliance Report, Office of Enforcement and Compliance Assurance, EPA document No. 305R13002. May 16, 2011, 94 p. http://www.epa.gov/comp liance/resources/reports/accomplishments/sdwa/sdwacom2011.pdf.

[4] According to the EPA, of roughly 156,600 public water systems, 142,400 rely on ground water and 14,200 rely on surface water. Among roughly 51,700 community water systems, 40,000 rely on ground water and 11,700 rely on surface water.

[5] The Energy Policy Act of 2005 (P.L. 109-58, §322) amended SDWA Section 1421(d) to specify that the definition of "underground injection" excludes the injection of fluids or propping agents (other than diesel fuels) used in hydraulic fracturing operations related to oil, gas, or geothermal production activities.

[6] The wells are classified as follows: Class I (inject hazardous wastes, industrial non-hazardous liquids, or municipal wastewater beneath the lowermost USDW); Class II (inject brines and other fluids associated with oil and gas production, and hydrocarbons for storage); Class III (inject fluids associated with solution mining of minerals beneath the lowermost USDW); Class IV (inject hazardous or radioactive wastes into or above USDWs and generally are banned); Class V (all injection wells not covered under other classes—many of these wells inject non-hazardous fluids into or above USDWs, and are typically shallow, on-site disposal systems), and Class VI (inject carbon dioxide (CO2) for long-term geologic sequestration to reduce atmospheric emissions of CO2 from industrial sources).

[7] See U.S. Environmental Protection Agency, Drinking Water Infrastructure Needs Survey and Assessment: Fifth Report to Congress, Office of Water, EPA 816-R-13-006, April 2013, 70 p., http://water.epa.gov/ grants_funding/dwsrf/ upload/epa816r13006.pdf.

[8] For more information, CRS Report RS22037, Drinking Water State Revolving Fund (DWSRF): Program Overview and Issues, by Mary Tiemann.

[9] 42 U.S.C. 300g-6. From 1986 through January 3, 2014, "lead free" under SDWA Section 1417(d) was defined to mean not more than 0.2% lead for solders and fluxes, and not more than 8% lead for pipes and pipe fittings. The 1996 SDWA amendments further provided that, for plumbing fittings and fixtures, "lead free" referred to plumbing fittings and fixtures in compliance with industry standards established under Section 1417(e).

[10] As amended in 2011, Section 1417(d) defines "lead free" to mean "(A) not containing more than 0.2% lead when used with respect to solder and flux (unchanged from existing law); and (2) not more than a weighted average of 0.25% lead when used with respect to the wetted surfaces of pipes, pipe fittings, plumbing fittings, and fixtures." P.L. 111-380 established a formula to calculate the weighted average lead content of a pipe, pipe fitting, plumbing fitting, or fixture.

[11] U.S. Environmental Protection Agency, Summary of the Reduction of Lead in Drinking Water Act and Frequently Asked Questions, EPA 815-S-13-001, October 2013, p. 5.

[12] For more information, see CRS Report INB00006, Lead-Free Requirements for Fire Hydrants, by Mary Tiemann.

[13] Colonias generally are described as unincorporated communities or housing developments on the U.S. side of the U.S.-Mexico border that lack some or all basic infrastructure including plumbing and public water and sewer.

[14] The Department of the Interior, Environment, and Related Agencies Appropriations Act, 2006, P.L. 109-54, Title II, August 2, 2005, 119 Stat. 530, provided "That for fiscal year 2006 and thereafter, State authority under section 302(a) of P.L. 104-182 shall remain in effect."

INDEX

D

E

I

J

L

M

N

O

P

Q

R

S

T

U

V

W